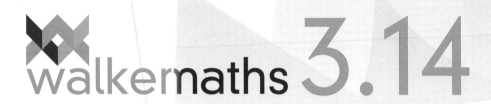

walkermaths 3.14

PROBABILITY DISTRIBUTIONS

NCEA Level 3 External

Second Edition

Charlotte Walker and Victoria Walker

Walker Maths 3.14 Probability Distributions 2ed
2nd Edition
Charlotte Walker
Victoria Walker

Designer: Cheryl Smith, Macarn Design
Production controller: Siew Han Ong

Any URLs contained in this publication were checked for currency during the production process. Note, however, that the publisher cannot vouch for the ongoing currency of URLs.

Acknowledgements
Cover photo courtesy of Shutterstock.

We wish to thank the Boards of Trustees of Darfield and Riccarton High Schools for allowing us to use materials and ideas developed while teaching. Our thanks also go to all past and present colleagues who have generously shared their expertise and ideas.

For product information and technology assistance,
in Australia call **1300 790 853**;
in New Zealand call **0800 449 725**

For permission to use material from this text or product, please email
aust.permissions@cengage.com

National Library of New Zealand Cataloguing-in-Publication Data
A catalogue record for this book is available from the National Library of New Zealand

978 0 17044693 8

Cengage Learning Australia
Level 7, 80 Dorcas Street
South Melbourne, Victoria Australia 3205

For learning solutions, visit **cengage.co.nz**

Printed in China by 1010 Printing International Limited.
10 11 12 26 25 24

CONTENTS

Remove and cut up.

 Formulae

These are the formulae that will supplied to you in the external examination.

Mean and variance of a discrete random variable	$\mu = E(X)$ $\quad = \Sigma x.P(X = x)$ $\sigma^2 = Var(X)$ $\sigma = SD(X)$ $\quad = \sqrt{\Sigma(x - \mu)^2 .P(X = x)}$ $\quad = \sqrt{E(X^2) - [E(X)]^2}$
Expectation algebra	$E[aX + b] = aE[X] + b$ $Var[aX + b] = a^2Var[X]$ $E[aX + bY] = aE[X] + bE[Y]$ $Var[aX + bY] = a^2Var[X] = b^2Var[Y]$ $\qquad\qquad$ if X, Y are independent
Continuous uniform distribution	The probability density function, $f(x)$, for a continous uniform distribution is defined as: $f(x) = \begin{cases} \dfrac{1}{b - a} & \text{for } a \le x \le b \\\\ 0, & \text{elsewhere} \end{cases}$
Standard normal distribution	$\left(Z = \dfrac{X - \mu}{\sigma}\right)$ 0 $\quad$ z Each entry gives the probability that the standardised normal random variable Z lies between 0 and z.
Binomial distribution	$P(X = x) = \binom{n}{x}\pi^x(1 - \pi)^{n-x}$ $\mu = n\pi, \qquad \sigma = \sqrt{n\pi(1 - \pi)}$
Poisson distribution	$P(X = x) = \dfrac{\lambda^x e^{-\lambda}}{x!}$ $\mu = \lambda, \qquad \sigma = \sqrt{\lambda}$

 ISBN: 9780170446938

Triangular distribution	The probability density function, $f(x)$, for a triangular distribution is defined as:
	$$f(x) = \begin{cases} 0, & x < a \\[2mm] \dfrac{2(x-a)}{(b-a)(c-a)} & a \le x \le c \\[2mm] \dfrac{2(b-x)}{(b-a)(b-c)} & c \le x \le b \\[2mm] 0, & x > b \end{cases}$$
	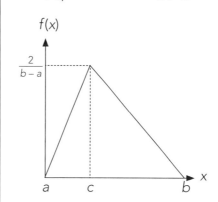
	Area of a triangle $= \dfrac{1}{2}$ base x height

 Glossary

Make your own glossary of key terms:

Term	Definition	Picture/Example
Normal distribution		
Binomial distribution		
Poisson distribution		
Uniform distribution		
Triangular distribution		
Mean		
Standard deviation		
Variance		
Random variable		
Parameter		

 ISBN: 9780170446938

What are probability distributions?

When data is collected from an experiment or an observational study, it can be modelled by a probability distribution. The probability of each outcome determines the shape of the distribution.

This standard will require you to investigate situations that involve elements of chance, and to use a range of probability distributions to solve problems.

Discrete vs continuous random variables

- A random variable is a variable whose values cannot be predicted, but are determined by the outcome of an experiment or an observational study.
- It is possible to allocate probabilities to each value that the random variable can take.
- Remember that for random variables:
  - upper case letters (e.g. X) represent the **name** of the random variable
  - lower case letters (e.g. x) represent the **values** it can take.

A descriptive (categorical) random variable is a variable in which:

- is usually a **word** or label, e.g. colour, breed of sheep
- **describes** something.

You will not study these in this standard.

A discrete random variable is a variable in which:

- values are usually **whole** numbers, e.g. the number of pets owned by students
- **counting** is involved
- it **is** possible to calculate a value for $P(X = x)$.

The discrete random variables that you will study for this standard:

- may not conform to any recognised pattern, but be defined by equations, a graph or a table
- may conform to the recognised characteristics of the binomial or Poisson distributions.

A continuous random variable is a variable in which:

- values are **measured** but not counted, e.g. students' heights
- $P(X = x) = 0$ for any value of x, e.g. you cannot be **exactly** 1.6 m (that means 1.6$\dot{0}$) tall
- you can find only the probability that x lies within some interval: $P(x_1 \leq X \leq x_2)$, e.g. you could find the probability that your height is **between** 1.60 and 1.61 m.

The continuous random variables that you will study for this standard are:

- the normal distribution
- the rectangular (continuous uniform) distribution
- the triangular distribution.

Complete the table by stating what type of variable each is.

Variable	Type of variable
Number of glow worms per m^2.	
Jellybean colour	
Length of foot	
Shoe size	
Time taken to complete a puzzle	
The cost of an insurance premium	
Volume of blood in an adult	
Distance between cities	
Names of the species of wasps living in New Zealand	
Density of wasp nests per hectare	
Make of car	
Mass of a car	
Car engine capacity	
Seating capacity of a car	
Total marks obtained in a quiz	
Achievement standard result	
Frequency of earthquakes	
Magnitude of an earthquake	
Height	
Adult clothing size (e.g. XL)	
Current flowing through a wire	
Quality of airline meal rated from 1 (poor) to 5 (excellent)	
Age in years	
Age in seconds	

Notice that the difference between discrete and continuous can blur when we consider very small units.

 ISBN: 9780170446938

The language of probability

You cannot begin to use probability distributions unless you can understand the notation and words used.

Let *X* be a **discrete** variable that can take values between 0 and 8.

Notation

Notation	The values that *X* can take are highlighted:	This could also be written as:
P(*X* = 3)	0 1 2 **3** 4 5 6 7 8	P(2 < *X* < 4)
P(*X* > 3)	0 1 2 3 **4 5 6 7 8**	P(*X* ≥ 4)
P(*X* < 3)	**0 1 2** 3 4 5 6 7 8	P(*X* ≤ 2)
P(*X* ≥ 3)	0 1 2 **3 4 5 6 7 8**	P(*X* > 2)
P(*X* ≤ 3)	**0 1 2 3** 4 5 6 7 8	P(*X* < 4)
P(2 < *X* < 7)	0 1 2 **3 4 5 6** 7 8	P(3 ≤ *X* ≤ 6)
P(2 ≤ *X* ≤ 7)	0 1 **2 3 4 5 6 7** 8	P(1 < *X* < 8)

Complete the table.

Notation	Highlight the numbers included in the description in the left column:	This could also be written as:
P(*X* > 5)	0 1 2 3 4 5 6 7 8	P(*X* ≥)
P(*X* ≤ 1)	0 1 2 3 4 5 6 7 8	P(*X* <)
P(0 < *X* < 4)	0 1 2 3 4 5 6 7 8	P(≤ *X* ≤)
P(*X* ≥ 7)	0 1 2 3 4 5 6 7 8	P(*X* >)
P(*X* < 6)	0 1 2 3 4 5 6 7 8	P(*X* ≤)
P(*X* = 5)	0 1 2 3 4 5 6 7 8	P(< *X* <)
P(1 ≤ *X* ≤ 6)	0 1 2 3 4 5 6 7 8	P(< *X* <)

Words

Possible wording	The values that X can take are highlighted:	Notation
P(X equals 5) P(X is the same as 5) P(X is exactly 5)	0 1 2 3 4 **5** 6 7 8	P(X = 5) P(4 < X < 6)
P(X is more than 5) P(X is greater than 5) P(X is over 5) P(X exceeds 5)	0 1 2 3 4 5 **6 7 8**	P(X > 5) P(X ≥ 6)
P(X is less than 5) P(X is fewer than 5) P(X is under 5)	**0 1 2 3 4** 5 6 7 8	P(X < 5) P(X ≤ 4)
P(X is greater than or equal to 5) P(X is at least 5) P(X is no(t) less than 5) P(X is 5 or more)	0 1 2 3 4 **5 6 7 8**	P(X ≥ 5) P(X > 4)
P(X is less than or equal to 5) P(X is 5 or less) P(X is no(t) more than 5) P(X is at most 5)	**0 1 2 3 4 5** 6 7 8	P(X ≤ 5) P(X < 6)
P(X is between 1 and 4)	0 1 **2 3** 4 5 6 7 8	P(1 < X < 4) P(2 ≤ X ≤ 3)
P(X is between 1 and 4 inclusive)	0 **1 2 3 4** 5 6 7 8	P(1 ≤ X ≤ 4) P(0 < X < 5)

Complete the table.

Words	Highlight the values that X can take:	Notation(s)
P(X is exactly 7)	0 1 2 3 4 5 6 7 8	
P(X is greater than 4)	0 1 2 3 4 5 6 7 8	
P(X is between 0 and 3)	0 1 2 3 4 5 6 7 8	
P(X is less than 2)	0 1 2 3 4 5 6 7 8	
P(X is at least 3)	0 1 2 3 4 5 6 7 8	

 ISBN: 9780170446938

P(X is between 4 and 7 inclusive)	0	1	2	3	4	5	6	7	8		
P(X is 6 or less)	0	1	2	3	4	5	6	7	8		
P(X is greater than or equal to 7)	0	1	2	3	4	5	6	7	8		
P(X is under 3)	0	1	2	3	4	5	6	7	8		
P(X is over 2)	0	1	2	3	4	5	6	7	8		
P(X is 4 or more)	0	1	2	3	4	5	6	7	8		
P(X is more than 1)	0	1	2	3	4	5	6	7	8		
P(X is not more than 6)	0	1	2	3	4	5	6	7	8		
P(X exceeds 2)	0	1	2	3	4	5	6	7	8		
P(X is not less than 6)	0	1	2	3	4	5	6	7	8		
P(X is less than or equal to 2)	0	1	2	3	4	5	6	7	8		
P(X is at most 1)	0	1	2	3	4	5	6	7	8		

True probability, experimental probability and theoretical (model) probability

True probability is almost always unknown. It is the **actual** probability that an event occurs.

Example: It is assumed that the probability of getting a head when a fair coin is tossed is 0.5. However, it is almost certainly not exactly 0.500000…. For instance, the extra weight of metal in the head, the way it is tossed, the surface on which it lands, etc., will affect the probability of getting a head.

Experimental probability is the probability obtained from an **experiment** or an **observational study**.

Example: If we toss a coin 50 times and get 22 heads, the experimental probability of a head is 0.44. The larger the number of trials, the closer the experimental probability will be to the true probability.

Theoretical (model) probability is the probability obtained from a **probability model**. Probability models are based on mathematical theory and observations about the behaviour of objects in an idealised world.

Example: The model estimate for the probability of a head when a fair coin is tossed is 0.5, on the assumption that both sides of the coin are equally likely to land on top.

Probability models in this standard
You have probably already studied the normal distribution model. Other models that you are likely to study this year are the:
- Poisson distribution
- binomial distribution
- rectangular (continous uniform) distribution
- triangular distribution.

Always remember:
- Models must always be applied in context.
- There are sometimes situations that no model fits.
- Sometimes more than one model may be appropriate.
- We can determine whether a particular model is a good fit for a situation only by conducting an experiment or an observational study.
- Models will hardly ever fit a situation perfectly.

 ISBN: 9780170446938

 Discrete random variables

1 General distributions

- These may be given In table or graph form.
- They may or may not conform to the recognised pattern of a specific distribution.
- You may be required to calculate the mean, standard deviation and/or variance.
- These formulae are on your formula sheet and you need to either know how to use them or be able to use your calculator to do the calculations for you.
- You may then be asked to use these values in further calculations.

Describing distributions

You need to:

1 state the highest and lowest values
2 state the mode
3 describe the shape.

Parameters: A distribution can also be described using its parameters. The most common parameters are the mean and standard deviation, but you will meet others.

For example, the parameters for the normal distribution are the mean (which identifies the middle of the distribution) and the standard deviation (which describes the spread).

Greek letters: Parameters are given letters of the Greek alphabet.

For example, mean = μ (pronounced *mew*)

$$= E(X)$$
$$= \text{the expected value of } X$$

Standard deviation = σ (pronounced *sigma*)

$$= SD(X)$$

Formulae:

$$\mu = E(X)$$
$$= \Sigma x . P(X = x)$$

$$\sigma^2 = Var(X)$$
$$\sigma = SD(X)$$
$$= \sqrt{\Sigma(x - \mu)^2 . P(X = x)}$$
$$= \sqrt{E(X^2) - [E(X)]^2}$$

Theoretical formula.

Formula that makes calculation easier.

Describing distributions

Features to include in your description:

1 The **highest** and **lowest** values that the distribution can take.
 These are the *x* values of the right and left columns of the graph.
2 The **mode(s)**.
 This is the value for *x* at the highest point on the graph.
3 The **shape**.

Normal distribution
(hill/mound shape, symmetrical, bell-shaped curve)

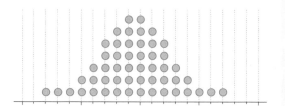

Right skew
(the tail is on the right-hand side)

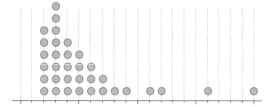

Left skew
(the tail is on the left-hand side)

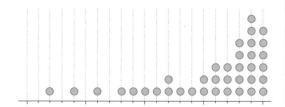

Bimodal
(there are two peaks)

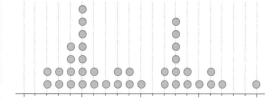

Uniform or rectangular
(looks like a box)

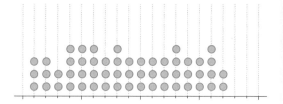

Triangular

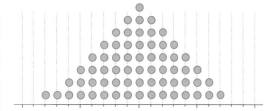

Irregular
(no real pattern)

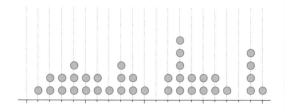

 ISBN: 9780170446938

Example: Describe the features of the following distribution.

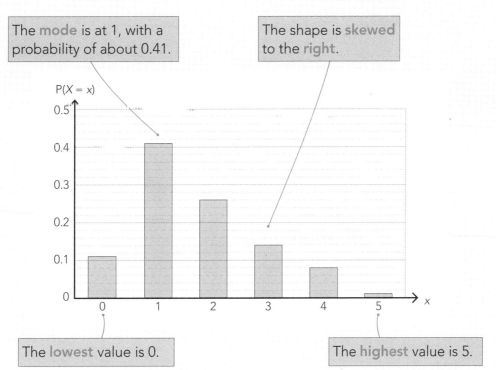

The **mode** is at 1, with a probability of about 0.41.

The shape is **skewed** to the **right**.

The **lowest** value is 0.

The **highest** value is 5.

Answer: The values range from 0 to 5, with a mode at 1. The probability that x takes the value 1 is about 0.41. The distribution is skewed to the right.

Describe the following distributions.

1

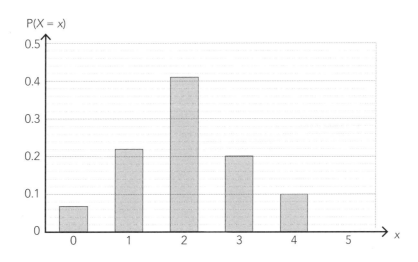

_____

_____

_____

ISBN: 9780170446938

2

n	0	1	2	3	4	5
P(N = n)	0.16	0.18	0.17	0.15	0.16	0.18

If it helps sketch the graph.

3

n	0	1	2	3	4	5	6
P(N = n)	0.04	0.09	0.14	0.21	0.38	0.12	0.02

4

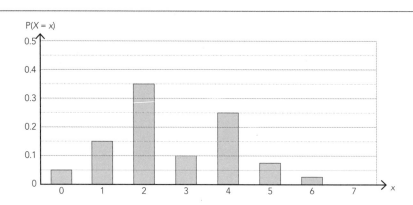

5

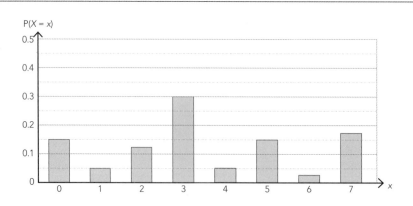

 ISBN: 9780170446938

Calculating the mean, standard deviation and variance of distributions

Example:

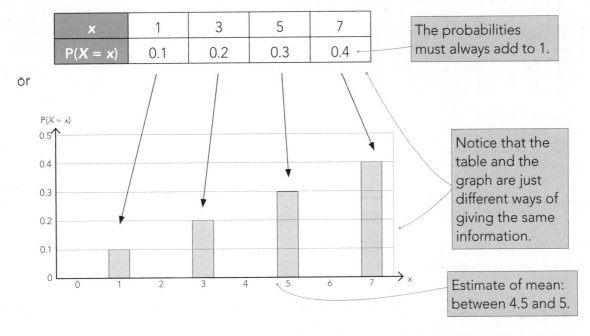

x	1	3	5	7
P(X = x)	0.1	0.2	0.3	0.4

The probabilities must always add to 1.

or

Notice that the table and the graph are just different ways of giving the same information.

Estimate of mean: between 4.5 and 5.

Before calculating the mean, estimate its value by looking for the middle of the shaded areas on the graph.

1 Using the formulae (these are all on your formula sheet)

$$\mu = E(X)$$
$$= \Sigma x.P(X = x)$$
$$= (1 \times 0.1) + (3 \times 0.2) + (5 \times 0.3) + (7 \times 0.4)$$
$$= 5$$

Check that this answer is close to your estimate.

$$\sigma = SD(X)$$
$$= \sqrt{E(X^2) - [E(X)]^2}$$
$$= \sqrt{[(1^2 \times 0.1) + (3^2 \times 0.2) + (5^2 \times 0.3) + (7^2 \times 0.4)] - 5^2}$$
$$= \sqrt{[0.1 + 1.8 + 7.5 + 19.6] - 25}$$
$$= \sqrt{29 - 25}$$
$$= 2$$

$$\sigma^2 = Var(X)$$
$$= 2^2$$
$$= 4$$

2 Using your calculator

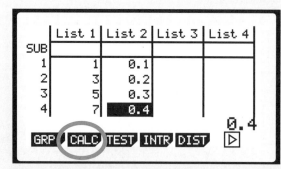

→ Menu

→ Stat

Input your data

→ CALC

→ 1Var

Make sure your settings are correct.

This needs to be on List 2. Remember, when your calculator is reset, this changes!

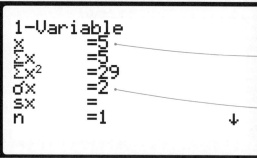

This is your mean, or $E(X)$.

This is your standard deviation, or $SD(X)$.

To find the variance, or $Var(X)$:

$$Var(X) = (SD(X))^2$$
$$= (2)^2$$
$$= 4$$

This is written $\sigma^2 = Var(X)$ on your formula sheet.

 ISBN: 9780170446938

For each of the following distributions, complete the table (if necessary) and then calculate the mean, standard deviation and variance.

1

x	0	1	2	3
P(X = x)	0.1	0.4		0.2

Estimate of mean = _____

$\mu =$ _____

_____

_____

$\sigma =$ _____

_____

_____

_____

$\sigma^2 =$ _____

2

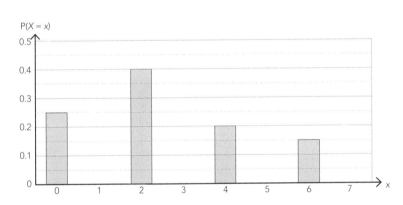

Estimate of mean = _____

$\mu =$ _____

_____

_____

$\sigma =$ _____

_____

_____

_____

$\sigma^2 =$ _____

3 Anika is a netball shooter. At the end of each practice, she takes five test shots and records the number of goals. After 20 practices, she has the following data. The random variable G represents the number of goals scored from the five test shots.

g	0	1	2	3	4	5
$P(G = g)$	0	0.05	0.25	0.4	0.2	

Estimate of mean = _____

_____

_____

_____

_____

_____

_____

_____

4 Tane had a die that he suspected was not throwing equal numbers of the numbers one to six. In order to test it, he threw it 100 times and recorded the result of each throw.

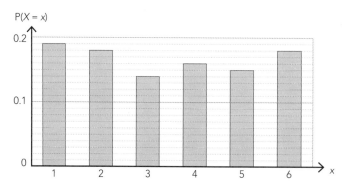

Estimate of mean = _____

_____

_____

_____

_____

_____

_____

_____

5 Let the random variable B represent the theoretical number of boys born in a family of four.

b	0	1	2	3	4
P(B = b)	0.0625	0.25	0.375	0.25	0.0625

Estimate of mean = _____

_____

_____

_____

_____

_____

_____

_____

6 Trevor has devised a 'fair' game. This means that the expected outcome, E(W), is 0. It costs $2 to enter. At the end of the game, the winner collects $16, and the runner-up gets $4. The table shows the expected 'winnings'.

w	$14	$2	-$2
P(W = w)	0.1	0.1	0.8

a Explain why the only possible outcomes are $14, $2 and –$2.

_____

_____

_____

b Perform a calculation to show that E(W) = 0.

_____

_____

_____

c Find the standard deviation SD(W) and variance Var(W).

_____

_____

_____

Linear combinations of a discrete random variable: expectation algebra

Linear combinations of a discrete random variable occur where:

1 A random variable is multiplied by a constant (a).

$$E(aX) = a \times E(X)$$

$$Var(aX) = a^2 \times Var(X)$$

$$SD(aX) = \sqrt{VAR(aX)}$$

> Note: the variance is derived from the **squares** of differences, so the constant **must** be squared.

> **Always** calculate the **variance first** and then use it to calculate the standard deviation:
> $$SD(X) = \sqrt{VAR(X)}$$

Example: Ella makes and sells cakes of soap which are packed in small boxes. She sells boxes containing a single soap for $2.50 and boxes which contain three cakes of soap for $7.00. The table shows the probability distribution for the amount of money (M) she earns from each sale:

M ($)	2.50	7.00
$P(M = m)$	0.64	0.36

For this distribution,

$$E(M) = \$4.12$$
$$Var(M) = 4.666$$
$$SD(M) = \$2.16$$

> If you have forgotten how to calculate these, look back to pages 17 and 18.

Her Dad thinks that she is selling her soaps too cheaply, and that she should increase her prices by 20%. Calculate the mean, variance and standard deviation for the money she would earn from each sale if she does as her Dad suggests.

$$E(1.2M) = 1.2 \times E(M)$$

$$= 1.2 \times \$4.12$$

$$= \$4.94$$

$$Var(1.2M) = 1.2^2 \times Var(M)$$

$$= 1.2^2 \times \$4.666$$

$$= 6.720$$

$$SD(1.2M) = \sqrt{Var(1.2M)}$$

$$= \sqrt{6.720}$$

$$= \$2.59$$

> Do not put units on variances: remember these are obtained by **squaring** differences.

ISBN: 9780170446938

2 A random variable has a constant (b) added to or subtracted from it.

$$E(X \pm b) = E(X) \pm b$$

$$Var(X \pm b) = Var(X) \pm Var(b)$$

$$= Var(X)$$

$$SD(X \pm b) = \sqrt{Var(X)}$$

b is a constant, so it does not vary. $\therefore$ Var(b) = 0.

Example: Ella's aunt suggests that it would be much simpler to add $0.50 to both her prices. Calculate the mean, variance and standard deviation for the money she would earn from each sale if she does as her Aunt suggests. (Remember E(M) = 4.12 and Var(M) = 4.666).

$$E(M + 0.50) = E(M) + \$0.50$$

$$= \$4.12 + \$0.50$$

$$= \$4.62$$

$$Var(M + 0.50) = Var(M)$$

$$= 4.666$$

$$SD(M + 0.50) = \sqrt{Var(M + 0.50)}$$

$$= \sqrt{4.666}$$

$$= \$2.16$$

$0.50 is always $0.50, so Var($0.50) = 0

Notice that the variance and the standard deviation are **not changed** by the addition of a constant.

3 A random variable is multiplied by a constant (a) and it has a constant (b) added to or subtracted from it.

$$E(aX + b) = a \times E(X) + b$$

$$Var(aX + b) = a^2 \times Var(X) + Var(b)$$

$$= a^2 \times Var(X)$$

$$SD(aX + b) = \sqrt{Var(aX + b)}$$

These two formulae are on your **formula sheet**.

Example: Ella's brother thinks that she should increase her prices by 10% and add $0.50 to each price. Calculate the mean, variance and standard deviation for the money she would earn from each sale if she does as her brother suggests.

$$E(1.1M + 0.50) = E(1.1M) + 0.50$$

$$= 1.1 \times 4.12 + 0.50$$

$$= \$5.03$$

$0.50 is always $0.50, so Var($0.50) = 0

$$Var(1.1M + 0.50) = 1.1^2 \times Var(M)$$

$$= 1.1^2 \times 4.666$$

$$= 5.646$$

$$SD(1.1M + 0.50) = \sqrt{Var(1.1M + 0.50)}$$

$$= \sqrt{5.647}$$

$$= \$2.38$$

ISBN: 9780170446938

Answer the following.

1 A random variable (X) has a mean of 11 and a variance of 4. Calculate the following:

a   SD(X) = _____

b   E(5X) = _____

c   Var(5X) = _____

d   SD(5X) = _____

e   E(X + 5) = _____

f   Var(X + 5) = _____

g   SD(X + 5) = _____

h   E(3X – 2) = _____

i   Var(3X – 2) = _____

j   SD(3X – 2) = _____

2 The school breakfast programme offers toast and a variety of spreads. The table shows the distribution of the number of pieces of toast eaten by each student.

t	0	1	2	3	4
P(T = t)	0.11	0.36	0.29	0.19	0.05

a   Calculate the mean and variance for number of pieces of toast (T) eaten by each student.

E(T) = _____   Var(T) = _____

b   It costs a total of 17 c (6 c for the bread and 11 c for spreads) for each piece of toast. Calculate the mean and variance for cost per student of toast with spreads.

_____

_____

c   Disposable plates cost 12 c per student. Every student has one plate, and plates are not reused. Calculate the mean and standard deviation for cost per student for a plate with toast and spreads.

_____

_____

3 As well as toast, the school breakfast programme offers Weet-bix and milk. The table shows the distribution of the number of Weet-bix eaten by each student.

w	0	1	2	3	4	5	6
P(W = w)	0.19	0.12	0.29	0.16	0.14	0.07	0.03

a   Calculate the mean and variance for number of Weet-bix (W) eaten by each student.

E(W) = _____   Var(W) = _____

b   Weet-bix cost 15 c each and each Weet-bix requires on average, 8 c worth of milk. The students eat from disposable bowls which cost 25 c each. Calculate the mean and standard deviation for cost per student for a Weet-bix breakfast.

_____

_____

4 Carl's Cabins hold a maximum of four people each. The table below shows the probability distribution for *P*, the number of people per night staying in each cabin.

p	0	1	2	3	4
P(P = p)	0.12	0.19	0.34	0.14	0.21

a Calculate the mean and variance for the number of people staying in each cabin.

E(*P*) = _____   _____   Var(*P*) = _____

b He charges a fixed nightly rate of $28 (regardless of how many people stay in it) for each cabin in order to cover cleaning, maintenance, etc. In addition he charges $18 per head per night. Calculate his average nightly income from each cabin.

_____

c Calculate the standard deviation for his nightly income from each cabin.

_____

_____

d How would the standard deviation for his nightly income from each cabin change if he increased the fixed nightly rate from $28 to $35? Explain your answer.

_____

_____

5 The graph shows the probability distribution (to 2 dp) for the number of cars a salesman sells in a week.

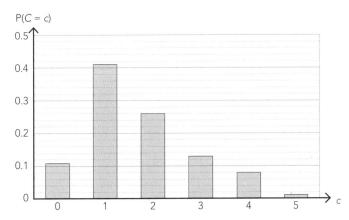

a Calculate the mean and variance for the number of cars he sells in a week.

E(*C*) = _____   Var(*C*) = _____

b If the mean value of each car sold is $23 750, calculate the expected total value of cars sold by this salesman in a week.

_____

c Calculate the standard deviation for the value of cars sold each week.

_____

_____

Linear combinations of several discrete distributions: expectation algebra

$E(aX + bY) = aE(X) + bE(Y)$

$Var(aX + bY) = a^2Var(X) + b^2Var(Y)$ **provided that X and Y are independent.**

$SD(aX + bY) = \sqrt{VAR(aX + bY)}$

Example:

Ella sells boxes containing a single soap (S) for \$2.50 and boxes that contain three cakes of soap (T) for \$7.00. She has kept records of the numbers of each that she sells at the Sunday market.

She knows:
$E(S) = 32.3$	$SD(S) = 5.6$
$E(T) = 18.1$	$SD(T) = 4.3$

Calculate the mean and standard deviation for the amount of money that she earns from these soaps at each Sunday market.

$E(2.5S + 7T) = 2.5E(S) + 7E(T)$
$= 2.5 \times 32.3 + 7 \times 18.1$
$= \$207.45$
$Var(2.5S + 7T) = 2.5^2 \times 5.6^2 + 7^2 \times 4.3^2$
$= 1102$
$SD(2.5S + 7T) = \sqrt{1102}$
$= \$33.20$

> To justify adding the variances, we must assume that the sales of single soaps are **independent** of the sales of boxes containing three soaps.

Answer the following questions.

1 A random variable (X) has a mean of 15 and a variance of 9.
A random variable (Y) has a mean of 7 and a variance of 4.
A random variable (Z) has a mean of 5 and a variance of 1.
X, Y and Z are independent of each other. Calculate the following.

a $E(X + Y) = $ _____

b $E(4X + Z) = $ _____

c $E(2X + 3Y + 4) = $ _____

d $Var(X + Y) = $ _____

e $SD(X + Y) = $ _____

f $Var(X + Z - 5) = $ _____

g $SD(3Y + 5) = $ _____

h $SD(2X + 4Z) = $ _____

i $Var(3X + 5Y - 2) = $ _____

j $SD(2Y + 5Z) = $ _____

2 The variances for two random variables X and Y are given below:

$Var(X) = 5.2$ $\qquad\qquad$ $Var(Y) = 3.6$

If $Var(X + Y) = 8.8$, are the variables X and Y independent? _____
Explain your answer.

_____

_____

3   The standard deviations for two random variables X and Y are given below:

     $SD(X) = 1.2$ $SD(Y) = 0.9$

If Var $(X + Y) = 2.1$, are the variables X and Y independent? _____
Justify your answer.

_____

_____

4   **a**   For the random variable P, $Var(P) = 5.264$.

     If $Var(P + Q) = 6.893$, and the random variables P and Q are independent, calculate $Var(Q)$ and explain your reasoning.

     _____

     _____

   b   Calculate the standard deviation of the random variable $3P + 2Q$.

     _____

     _____

5   The table below shows the probability distribution for the random variable T.

t	0	1	2	3	4
P(T = t)	0.15	0.35	0.30	0.15	0.05

   a   Calculate the mean and standard deviation for T.

     $E(T) =$ _____ $SD(T) =$ _____

   b   For a second random variable S, $Var(S) = 1.793$.

     If $Var(S + T) = 2.95$, are the variables S and T independent? Show your reasoning and comment on your answer.

     _____

     _____

   c   Estimate the standard deviation of the random variable $(0.5S + 1.2T)$. Support your answer with appropriate statistical statements.

     _____

     _____

   d   Explain why you can only estimate the standard deviation of $(0.5S + 1.2T)$.

     _____

     _____

A tricky distinction:

Example: Ella also sells candles for $4.00 each at the market. Let the random variable N represent the number of candles she sells at a typical Sunday market.

$$E(N) = 12.4 \qquad\qquad SD(N) = 2.8$$

Find the mean and standard deviation for the value of the candles that she sells at a typical Sunday market.

In this case, the random variable ($4N$) is a multiple of **another random variable (N):** every **candle sells for** $4, **so the value of the candles sold is** four times **the number sold (N).**

$$E(4N) = 4 \times 12.4$$
$$= \$49.60$$
$$Var(4N) = Var(\$N)$$
$$= 4^2 \times 2.8^2$$
$$= 125.44$$
$$SD(4N) = \sqrt{125.44}$$
$$= \$11.20$$

Where **every** candle has a value of $4, so the multiplier (4) **must be squared**.

Compare this with the following:

Example: As in the example above, let the random variable N represent the number of candles she sells at a typical Sunday market.

$$E(N) = 12.4 \qquad\qquad SD(N) = 2.8$$

Calculate the mean and standard deviation for the number of candles that she sells in the month of March, during which she sells candles at four Sunday markets.

In this case, the random variable is the sum of **single random variables, each of which has its own mean and standard deviation: on each of the four Sundays she can be expected to sell 12.4 candles, and on each of the four Sundays the standard deviation for the number sold will be 2.8.**

$$E(4N) = E(N + N + N + N)$$
$$= 12.4 + 12.4 + 12.4 + 12.4$$
$$= 4 \times 12.4$$
$$= \$49.60$$
$$Var(4N) = Var(N + N + N + N)$$
$$= Var(N) + Var(N) + Var(N) + Var(N)$$
$$= 4 \times 2.8^2$$
$$= 31.36$$
$$SD(4N) = \sqrt{31.36}$$
$$= \$5.60$$

For the **sum** of the number sold at four markets, the multiplier is **not** squared.

   ISBN: 9780170446938

6 **a** Ella's friend sells pumpkins for $3 each at the Sunday market. Let the random variable P represent the number of pumpkins that she sells.

$$E(P) = 7.28 \qquad\qquad SD(P) = 1.87$$

Calculate the mean and standard deviation for the money that she earns from selling pumpkins.

_____

_____   _____

_____

b Calculate the mean and the standard deviation for the number of pumpkins that she sells in total at the four Sunday markets during March.

_____

_____

_____

c She also sells melons for $5 each. Let the random variable M represent the number of melons that she sells.

$$E(M) = 15.07 \qquad\qquad SD(M) = 3.97$$

Calculate the mean and standard deviation for the random variable V, which represents the **total value of pumpkins and melons** sold at a Sunday market. Support your answer with appropriate statistical statements and calculations.

_____

_____

_____

d Calculate the mean and standard deviation for the random variable T, which represents the total number of pumpkins and melons sold at the four Sunday markets during March. Support your answer with appropriate statistical statements and calculations.

_____

_____

_____

e State any assumption you have made in parts **c** and **d**, and discuss its validity in this case.

_____

_____

_____

7 For the school breakfast programme, T represents the number of pieces of toast eaten by each student each day.

$$E(T) = 1.71 \qquad\qquad Var(T) = 1.106$$

On Monday, 49 students turn up wanting breakfast. If 38 students want toast, calculate the mean and standard deviation for the total number of pieces of toast that will be needed.

_____

_____

_____

8 **a** The students need plates and bowls for their breakfasts. Let P represent the number of plates and B represent the number of bowls that are used on a typical day.

$$E(P) = 35.6 \qquad\qquad SD(P) = 4.1$$
$$E(B) = 24.3 \qquad\qquad SD(B) = 3.7$$

Plates cost 12c each and bowls cost 25c each. Calculate the mean and standard deviation for the cost of bowls and plates that are used on a typical day. Show appropriate calculations, and justify your reasoning.

_____

_____

_____

_____

_____

b Calculate the mean and standard deviation for the total number of plates and bowls (T) used in a term that has 49 school days. Show appropriate calculations, and justify your reasoning.

_____

_____

_____

_____

_____

c Give a reason why the reasoning that you used in parts **a** and **b** may not be valid.

_____

_____

_____

2   The Poisson distribution

Characteristics of the Poisson distribution:
- It is the distribution of **rare** events.
- It is often recognised by **rates** and the word '**per**' or '**every**', e.g. number of earthquakes **per** year.
- Lower limit is **0**, but usually there **is no upper limit**.
- There must be **discrete** events that occur in a **continuous** but finite interval of time or space.

For example, number of accidents **per** kilometre of road.

Discrete event   Finite interval of length

Conditions that must meet for the Poisson distribution to be used:
1   Each occurrence is **independent** of others.
2   Events must **not occur simultaneously**.
3   Events must occur **randomly** and **unpredictably**.
4   For a small interval, the probability of an event occurring is **proportional to the size of the interval**.

Parameter for the Poisson distribution: λ (*lambda*) = the mean

Formula for calculating P(X = x):

$$P(X = x) = \frac{\lambda^x e^{-\lambda}}{x!}$$

Mean, variance and standard deviation for the Poisson distribution:

Mean = λ
Variance = λ
Standard deviation = $\sqrt{\lambda}$

A distinctive feature of the Poisson distribution is that theoretically **mean = variance**.

Examples of graphs of theoretical Poisson distributions:

$\lambda = 0.5$ $\lambda = 1.5$

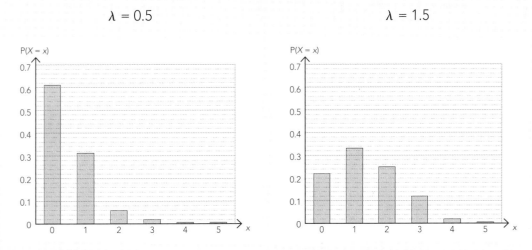

These tend to be **skewed to the right** because events are rare.

ISBN: 9780170446938

Example 1: A school has on average six broken windows per term. Calculate the probability that it will get four broken windows in a term.

$$\lambda = 6$$
$$x = 4$$

1 Using the formula:

$$P(X = x) = \frac{\lambda^x e^{-\lambda}}{x!}$$

$$= \frac{6^4 \times e^{-6}}{4!}$$

$$P(X = 4) = 0.1339$$

2 Using your calculator:

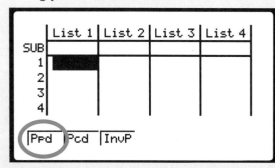

$\longrightarrow$ **Menu**
$\longrightarrow$ **Stat**
$\longrightarrow$ **Dist**
$\longrightarrow$ **POISN**
$\longrightarrow$ **Ppd** — Poisson **p**oint **d**istribution (see page 30)

Make sure the setting is Var, not List.
Note: This will revert back to List when your calculator is reset.

Remember the mean is *lambda*.

$$P(X = 4) = 0.1339$$

3 Using tables (see back of book):

Values for lambda

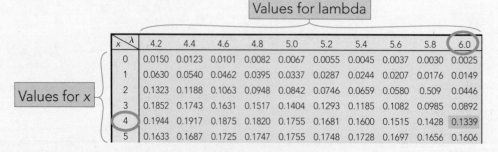

x \ λ	4.2	4.4	4.6	4.8	5.0	5.2	5.4	5.6	5.8	6.0
0	0.0150	0.0123	0.0101	0.0082	0.0067	0.0055	0.0045	0.0037	0.0030	0.0025
1	0.0630	0.0540	0.0462	0.0395	0.0337	0.0287	0.0244	0.0207	0.0176	0.0149
2	0.1323	0.1188	0.1063	0.0948	0.0842	0.0746	0.0659	0.0580	0.509	0.0446
3	0.1852	0.1743	0.1631	0.1517	0.1404	0.1293	0.1185	0.1082	0.0985	0.0892
4	0.1944	0.1917	0.1875	0.1820	0.1755	0.1681	0.1600	0.1515	0.1428	0.1339
5	0.1633	0.1687	0.1725	0.1747	0.1755	0.1748	0.1728	0.1697	0.1656	0.1606

Values for x

$$P(X = 4) = 0.1339$$

Answer in a sentence:
The probability that there will be four broken windows in a term is 0.1339.

Example 2: The mean number of stoats in a reserve is 2.4 per hectare. Calculate the probability that there are more than 3 in a hectare.

$$\lambda = 2.4$$
$$x = 4, 5, 6, 7, 8, 9, 10, \ldots$$

> Notice that, although the distribution is discrete, the average number (λ) is not usually discrete.

1 Using the formula:

$$P(X = x) = \frac{\lambda^x e^{-\lambda}}{x!}$$

$$P(X > 3) = \frac{2.4^4 \times e^{-2.4}}{4!} + \frac{2.4^5 \times e^{-2.4}}{5!} + \frac{2.4^6 \times e^{-2.4}}{6!} + \frac{2.4^7 \times e^{-2.4}}{7!}$$

> This can be tedious, so is not recommended!

$$+ \frac{2.4^8 \times e^{-2.4}}{8!} + \frac{2.4^9 \times e^{-2.4}}{9!} + \frac{2.4^{10} \times e^{-2.4}}{10!} + \ldots$$

> You need to calculate these until the last term is so small it makes no difference. $P(X = 10) = 0.0002$

$$P(X = 4, 5, 6, 7, 8, 9, 10) = 0.2213$$

2 Using your calculator:

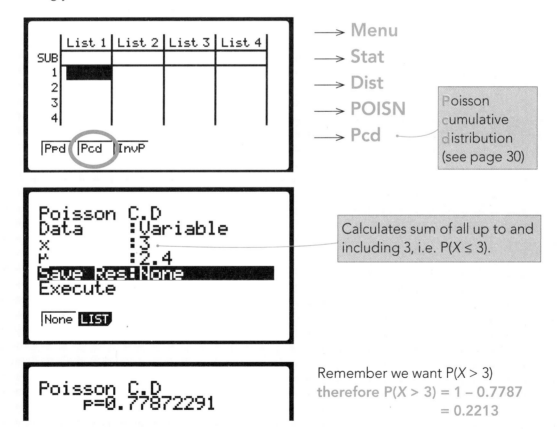

$\longrightarrow$ **Menu**

$\longrightarrow$ **Stat**

$\longrightarrow$ **Dist**

$\longrightarrow$ **POISN**

$\longrightarrow$ **Pcd** — Poisson cumulative distribution (see page 30)

> Calculates sum of all up to and including 3, i.e. $P(X \leq 3)$.

Remember we want $P(X > 3)$
therefore $P(X > 3) = 1 - 0.7787$
$= 0.2213$

3 Using tables:

Either: Find the probability that $x = 4, 5, 6, 7, 8, 9, 10, \ldots$

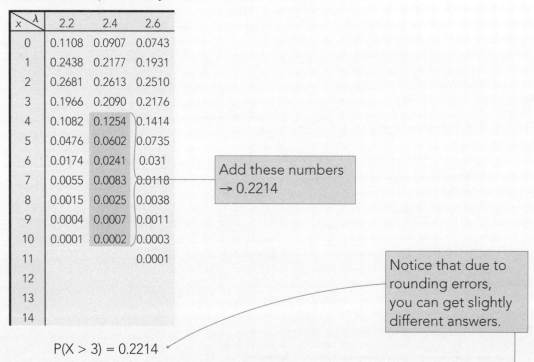

x \ λ	2.2	2.4	2.6
0	0.1108	0.0907	0.0743
1	0.2438	0.2177	0.1931
2	0.2681	0.2613	0.2510
3	0.1966	0.2090	0.2176
4	0.1082	0.1254	0.1414
5	0.0476	0.0602	0.0735
6	0.0174	0.0241	0.031
7	0.0055	0.0083	0.0118
8	0.0015	0.0025	0.0038
9	0.0004	0.0007	0.0011
10	0.0001	0.0002	0.0003
11			0.0001
12			
13			
14			

Add these numbers → 0.2214

Notice that due to rounding errors, you can get slightly different answers.

$$P(X > 3) = 0.2214$$

Or: Find the probability that $x = 0, 1, 2$ or 3, then subtract your answer from 1.

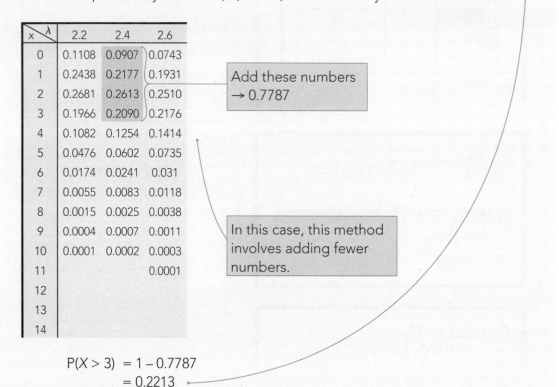

x \ λ	2.2	2.4	2.6
0	0.1108	0.0907	0.0743
1	0.2438	0.2177	0.1931
2	0.2681	0.2613	0.2510
3	0.1966	0.2090	0.2176
4	0.1082	0.1254	0.1414
5	0.0476	0.0602	0.0735
6	0.0174	0.0241	0.031
7	0.0055	0.0083	0.0118
8	0.0015	0.0025	0.0038
9	0.0004	0.0007	0.0011
10	0.0001	0.0002	0.0003
11			0.0001
12			
13			
14			

Add these numbers → 0.7787

In this case, this method involves adding fewer numbers.

$$\begin{aligned} P(X > 3) &= 1 - 0.7787 \\ &= 0.2213 \end{aligned}$$

Answer in a sentence:
The probability that there will be more than 3 stoats in a hectare is 0.2214 (or 0.2213).

 ISBN: 9780170446938

Use of Ppd and Pcd on your calculator. Note: The values used in these graphs do not correspond to any real value for lambda.

Ppd finds P(X = x)

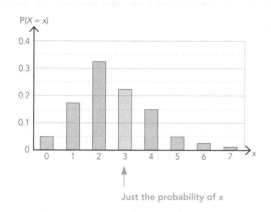

Just the probability of x

$P(X = 3) = 0.23$

Pcd finds P(X ≤ x)

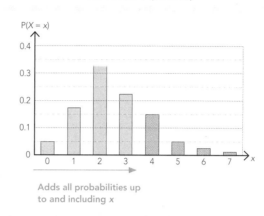

Adds all probabilities up
to and including x

$P(X \le 3) = P(X = 0) + P(X = 1) + P(X = 2)$
$+ P(X = 3)$
$= 0.05 + 0.18 + 0.32 + 0.23$
$= 0.78$

Calculations using Ppd and Pcd

Notation	The values that X can take are highlighted:	On a graphics calculator we would have to use:	x value for your calculator
$P(X = 3)$	0 1 2 **3** 4 5 6 7 8	Ppd	3
$P(X \le 3)$	0 1 2 3 4 5 6 7 8	Pcd	3
$P(X > 3)$	0 1 2 3 4 5 6 7 8	Pcd but $1 - P(X \le 3)$	3
$P(X < 3)$	0 1 2 3 4 5 6 7 8	Pcd but $P(X \le 2)$	2
$P(X \ge 3)$	0 1 2 3 4 5 6 7 8	Pcd but $1 - P(X \le 2)$	2

Complete the table using your calculator and $\lambda = 1.5$.

Notation	The values that X can take can be highlighted:	Ppd or Pcd?	x value for your calculator	Answer
$P(X = 4)$	0 1 2 3 4 5 6 7 8		4	
$P(X \le 1)$	0 1 2 3 4 5 6 7 8	Pcd		
$P(X > 2)$	0 1 2 3 4 5 6 7 8			$1 - P(X \le 2)$ $= 0.1912$
$P(X < 3)$	0 1 2 3 4 5 6 7 8		2	
$P(X \ge 4)$	0 1 2 3 4 5 6 7 8			

ISBN: 9780170446938

Answer the following questions.

1 Consider a Poisson distribution with $\lambda = 0.9$.
Complete the table.

	Values *x* can take	Probability
P(x = 0)	0 1 2 3 4 5 6 7 8	0.4066
P(x = 3)	0 1 2 3 4 5 6 7 8	
P(x = 5)	0 1 2 3 4 5 6 7 8	
P(x < 2)	0 1 2 3 4 5 6 7 8	
P(x > 1)	0 1 2 3 4 5 6 7 8	
P(x ≥ 3)	0 1 2 **3 4 5 6 7 8**	
P(1 < x < 4)	0 1 2 3 4 5 6 7 8	
P(x ≤ 4)	0 1 2 3 4 5 6 7 8	
P(4 < x ≤ 7)	0 1 2 3 4 5 6 7 8	

What do you notice about your last two answers? Explain your answer.

_____

2 Consider a Poisson distribution with $\lambda = 2.2$.
Complete the table.

	Values *x* can take	Probability
P(x is less than 1)		0.1108
P(x is between 2 and 4)		
P(x is exactly 10)		
P(x is no more than 2)		
P(x is at least 7)	7, 8, 9, 10, ...	
P(x is between 4 and 7)		
P(x is at least 6 and less than 9)		
P(x is not less than 2)		
P(x is at most 1)		

What is the mean of this distribution.

_____

Calculate the standard deviation of this distribution.

_____

PHOTOCOPYING OF THIS PAGE IS RESTRICTED UNDER LAW.   ISBN: 9780170446938

Applications of the Poisson distribution

Probability reminder – provided the events are independent

Probability of b and c	$P(b) \times P(c)$
Probability of b or c	$P(b) + P(c)$

Example: A factory produces climbing rope with a mean number of faults of 0.3 per 100 m.

a Calculate the theoretical probability that neither of two separate 100 m lengths of rope will have any faults.
Neither has any faults ⇒ one doesn't have any faults and the other doesn't have any faults.

$$\lambda \text{ for 100 m} = 0.3$$
$$P(X = 0) = 0.7408$$
∴ For no faults in each of two lengths, $P(X = 0) = 0.7408^2$
$$= 0.5488$$

b The factory produces a 100 m length of rope. Calculate the theoretical probability that it has one or two faults.

$$\lambda \text{ for 100 m} = 0.3$$
$$P(X = 1) = 0.2222$$
$$P(X = 2) = 0.0333$$
∴ $P(1 \text{ or 2 faults}) = 0.2222 + 0.0333$
$$= 0.2555$$

Rates in Poisson problems
- Remember the fourth condition: For a small interval, the probability of an event occurring is proportional to the size of the interval.
- In some questions you may be asked for a probability for an interval which is not the same as the interval in the question.

Example: Another factory produces climbing rope with a mean number of faults of 0.4 per 100 m.

a Calculate the theoretical probability that a 200 m length of rope will have no faults.

$$\lambda \text{ for 100 m} = 0.4$$
∴ $\lambda \text{ for 200 m} = 2 \times 0.4$
$$= 0.8$$
$$P(X = 0) = 0.4493$$

b The factory produces a 200 m length and a 50 m length. Calculate the theoretical probability that they both have exactly one fault each.

$$\lambda \text{ for 200 m} = 0.8 \rightarrow P(X = 1) = 0.3595$$
$$\lambda \text{ for } 50 \text{ m} = 0.2 \rightarrow P(X = 1) = 0.1637$$

∴ $P(\text{both have one fault}) = 0.3595 \times 0.1637$
$$= 0.0589$$

1 The mean number of cyclones striking a Pacific island is 0.6 per year.

a What is the theoretical probability that no cyclones strike the island during a year?

$\lambda =$ _____ $x =$ _____

_____

b What is the theoretical probability that two or three cyclones strike the island during a year?

$\lambda =$ _____ $x =$ _____

_____

c What is the theoretical probability that four cyclones or more strike the island during a year?

$\lambda =$ _____ $x =$ _____

_____

d What is the theoretical probability that fewer than two cyclones strike the island in a year?

$\lambda =$ _____ $x =$ _____

_____

e Write down the standard deviation and variance for this distribution.

$\sigma =$ _____ $\sigma^2 =$ _____

f On average, how many cyclones would you expect to strike the island in a 10-year period?

_____

_____

g What is the theoretical probability that five cyclones strike the island in a 10-year period?

$\lambda =$ _____ $x =$ _____

_____

h Write down four reasons why you think that the Poisson distribution is appropriate for this situation. Explain your reasons.

_____

_____

_____

_____

_____

2 Each year, this Pacific island experiences on average 1.2 earthquakes that are above four on the Richter scale.

a What is the theoretical probability that there is one earthquake that is over four on the Richter scale on the island during a year?

$\lambda =$ _____ $x =$ _____

_____

b What is the theoretical probability that there are three or four earthquakes that are over four on the Richter scale on the island during a year?

$\lambda =$ _____ $x =$ _____

_____

c What is the theoretical probability that there are five or fewer earthquakes that are over four on the Richter scale on the island during a year?

$\lambda =$ _____ $x =$ _____

_____

d What is the theoretical probability that there are more than two earthquakes that are over four on the Richter scale on the island during a year?

$\lambda =$ _____ $x =$ _____

_____

e What is the theoretical probability that there are no fewer than three earthquakes that are over four on the Richter scale on the island during a year?

$\lambda =$ _____ $x =$ _____

_____

f On average, how many earthquakes that are above four on the Richter scale would you expect to occur during a five-year period?

_____

g What is the theoretical probability that this island will have one earthquake and one cyclone in a year?

_____

_____

h Write down a reason why the Poisson distribution may not be appropriate for this situation. Explain your answer.

_____

_____

3   On average a delicatessen sells one jar of pickled snails every two weeks.

a   What is the theoretical probability that the delicatessen sells three jars of pickled snails in a two-week period?

_____

_____

b   What is the theoretical probability that the delicatessen sells at least three jars of pickled snails in a two-week period?

_____

_____

c   What is the theoretical probability that the delicatessen sells no more than two jars (inclusive) of pickled snails in a two-week period?

_____

_____

d   Explain how your answers to **b** and **c** relate to each other.

_____

_____

e   On average, how many jars of pickled snails would you expect the delicatessen to sell in an eight-week period?

_____

f   What is the theoretical probability that the delicatessen sells 10 jars in an eight-week period?

_____

_____

g   What is the theoretical probability that the delicatessen sells fewer than three jars in an eight-week period?

_____

_____

h   There are four conditions for the Poisson distribution. List these, and explain whether or not these apply to this situation.

_____

_____

_____

_____

4 The delicatessen also sells pots of jellied eel. On average, they sell three pots per week.

a Use your tables or calculator to help you complete the graph of the theoretical probability distribution for the numbers of pots of jellied eels sold.

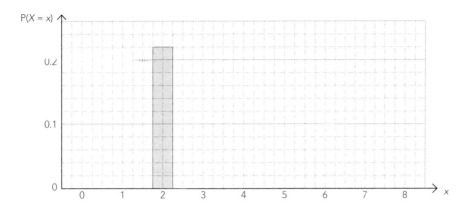

b Write down two characteristics of the Poisson distribution that your graph illustrates. Explain how these relate to the context.

_____

_____

c What is the probability that they sell at most two pots of jellied eels in a week?

_____

_____

d What is the probability that they sell no fewer than two pots of jellied eels in a week?

_____

_____

e On average, how many pots of jellied eels would you expect them to sell in two weeks?

_____

f What is the probability that they sell more than eight pots in a two-week period?

_____

_____

g Calculate the probability that they don't sell any jars of pickled snails or jellied eels in a week.

_____

_____

5 The graph shows the observed (experimental) probability distribution for the number of weta found on each square metre of a cave wall.

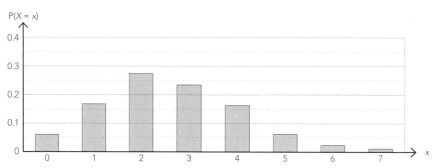

a Calculate the mean number of weta per square metre.

_____

_____

b Use your tables and an appropriate value of *lambda* to find the theoretical probabilities for each value of *X*. Show these on the graph above.

_____

c Use this value for *lambda* (from **b**) to find the theoretical probability that there are two or three weta in one square metre.

_____

_____

d Two separate areas, each of one square metre, are searched for weta. Using the value for *lambda* from **b**, find the theoretical probability that there were no weta in both squares.

_____

_____

e An area of 2 square metres was searched. Calculate the theoretical probability that no more than four weta were found.

_____

_____

f Compare the features of the graph of the actual distribution of weta with the theoretical Poisson distribution. Use these to justify whether or not a Poisson distribution is a good model for the number of weta found on each square metre of cave wall.

_____

_____

_____

_____

 ISBN: 9780170446938

6 The table shows the probability distribution for the number of wasp nests found in each hectare of a large section of bush in a national park.

n	0	1	2	3	4	5	6
P($N = n$)	0.51	0.16	0.10	0.08	0.06	0.05	0.04
			0.2303				

Experimental probabilities.

Theoretical Poisson probabilities.

a What is the experimental probability that there is at most one wasp nest per hectare in this section of the national park?

b What is the experimental probability that there is at least one but fewer than four wasp nests per hectare in this section of the national park?

c Use the table to justify why the Poisson distribution might be appropriate to describe this distribution of wasp nests.

d Calculate the mean and standard deviation for the number of wasp nests per hectare.

e Use the mean calculated in **d**, along with the tables or your calculator, to complete the table to show the theoretical Poisson distribution of wasp nests.

f Discuss whether a Poisson distribution is a good model for this distribution of wasp nests in the national park. As part of your discussion, you should compare the features of the observed (experimental) and model (theoretical) distributions.

Inverse Poisson problems

- In inverse problems you are **given a probability** and you use this to **calculate the parameter(s)**.
- Inverse Poisson problems can be done **only** if you know the value of $P(X = 0)$.
- Unless the P(0) matches a value in the tables, they **must** be done by **using the formula**.

Example 1: The probability that there are no faults in a 100 m length of climbing rope is 0.1496. Calculate the average number of faults per 100 m.

1 Using tables:

x \ λ	0.1	0.2	0.3	0.4	0.5	0.6	0.7	0.8	0.9	1.0
0	0.9048	0.8187	0.7408	0.6703	0.6065	0.5488	0.4966	0.4493	0.4066	0.3679
1	0.0905	0.1637	0.2222	0.2681	0.3033	0.3293	0.3476	0.3595	0.3659	0.3679
2	0.0045	0.0164	0.0333	0.0536	0.0758	0.0988	0.1217	0.1438	0.1647	0.1839
3	0.0002	0.0011	0.0033	0.0072	0.0126	0.0198	0.0284	0.03838	0.0494	0.0613
4		0.0001	0.0003	0.0007	0.0016	0.0030	0.0050	0.0077	0.0111	0.0153
5				0.0001	0.0002	0.0004	0.0007	0.0012	0.0020	0.0031
6							0.0001	0.0002	0.0003	0.0005
7										0.0001

x \ λ	1.1	1.2	1.3	1.4	1.5	1.6	1.7	1.8	1.9	2.0
0	0.3329	0.3012	0.2824	0.2466	0.2231	0.2019	0.1827	0.1653	0.1496	0.1353
1	0.3662	0.3614	0.3543	0.3452	0.3349	0.3230	0.3106	0.2975	0.2842	0.2707
2	0.2014	0.2169	0.2303	.2417	0.2510	0.2584	0.2640	0.2678	0.2700	0.2707

Check across the row for $X = 0$ until you find the correct value.

$$\therefore \lambda = 1.9$$

2 Using the formula:

$$P(X = x) = \frac{\lambda^x e^{-\lambda}}{x!}$$

But $\lambda^0 = 1$ and $0! = 1$

$$P(X = 0) = 0.1496 = \frac{\lambda^0 e^{-\lambda}}{0!}$$

$$\therefore 0.1496 = e^{-\lambda}$$

$$\lambda = -\log_e 0.1496$$

$$\lambda = 1.9$$

$\log_e$ = ln on your calculator.

Example 2: The probability that at least one fault occurs in a 100 m length of climbing rope is 0.8946. Calculate the average number of faults per 100 m.

$$P(\text{at least one fault}) = 0.8946$$
$$P(X = 0) = 1 - 0.8946$$
$$= 0.1054$$

1 Using tables:

x \ λ	2.2	2.4
0	0.1108	0.0907
1	0.2438	0.2177

0.1054 is between these values so the formula must be used.

2 Using the formula:

$$P(X = 0) = 0.1054 = \frac{\lambda^0 e^{-\lambda}}{0!}$$

$$\lambda = -\log_e 0.1054$$

$$\lambda = 2.25$$

Check: The tables showed that λ was between 2.2 and 2.4. $\therefore$ 2.25 is a reasonable answer.

ISBN: 9780170446938

Answer the following.

1 On a different Pacific Island, the probability that there are no cyclones in a year is 0.4493. Calculate the mean number of cyclones that strike the island in a year.

$P(X = 0) = $ _____

_____

_____

2 The probability that this Pacific island experiences no earthquakes that are above four on the Richter scale is 0.2698. For this island, calculate the mean number of earthquakes that are above four on the Richter scale each year.

$P(X = 0) = $ _____

_____

_____

3 If the neighbouring delicatessen sells no jars of pickled snails in 56% of weeks, calculate the mean number of jars sold each week.

_____

_____

_____

4 The neighbouring delicatessen sells at least one pot of jellied eel in 35% of weeks. Calculate the mean number of pots sold during a two-week period.

_____

_____

_____

5 In a cave, a biologist examined many separate sections of wall. Each was one square metre. She found at least one weta on 64% of these sections. Calculate the probability of finding no more than three weta in a four-square-metre section.

_____

_____

_____

6 The table shows the probability distribution for the number of wasp nests per hectare of bush. Calculate the probability that there are no fewer than two in a five-hectare block.

n	0	1	2	3	4
$P(N = n)$	0.644	0.284	0.062	0.009	0.001

_____

_____

_____

3 The binomial distribution

Characteristics of the binomial distribution:
- It is a **discrete** distribution.
- There are just **two outcomes**.

For example, throwing a six or a 'not six' when using dice.

Conditions that must be met for the binomial distribution to be used:
1 There must be only **two outcomes**.
2 The probability of success (π) must stay the same.
3 The number of trials (n) must be fixed $\Rightarrow$ lower limit of **0** and upper limit of **n**.
4 Each occurrence is **independent** of others.

Parameters for the binomial distribution: π or p = probability of one event
n = number of trials

Formula for calculating P(X = x):

$$P(X = x) = {}^nC_x p^x (1 - p)^{n-x}$$

On your formula sheet it looks like this:

$$P(X = x) = \binom{n}{x}\pi^x(1 - \pi)^{n-x}$$

Mean, variance and standard deviation for the binomial distribution:

Mean = np or $n\pi$

Variance = $np(1 - p)$ or $n\pi(1 - \pi)$

Standard deviation = $\sqrt{np(1 - p)}$ or $\sqrt{n\pi(1 - \pi)}$

Graphs of theoretical binomial distributions for n = 5:

$p = 0.1$ $p = 0.5$ $p = 0.9$

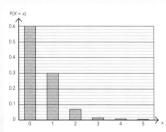 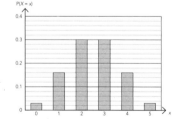 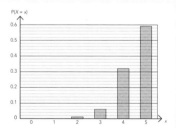

- This could to be **skewed to the right**, it could be **symmetrical** or **skewed to the left**, depending on the size of p.
- It is always unimodal.

Example 1: 25% of the bones in your body are in your feet. If five bones from your body are selected at random, what is the probability that three are from your feet?

$$p = 0.25$$
$$n = 5$$
$$x = 3$$

1 Using the formula:

$$P(X = 3) = {}^nC_x p^x (1 - p)^{n-x}$$
$$= {}^5C_3 (0.25)^3 (1 - 0.25)^{(5-3)}$$
$$= {}^5C_3 (0.25)^3 (0.75)^2$$
$$P(X = 3) = 0.0879$$

2 Using your calculator:

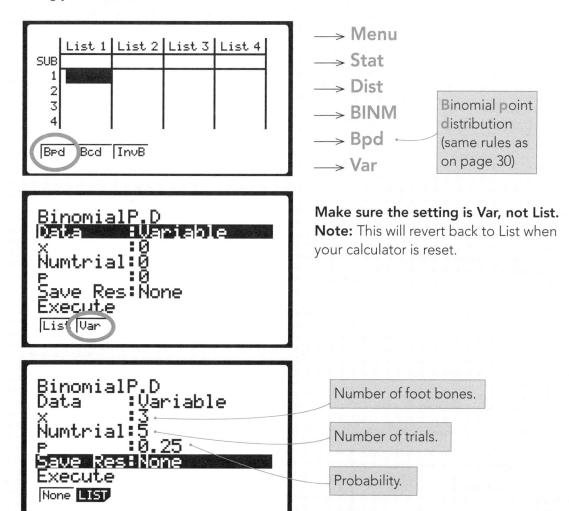

$\longrightarrow$ Menu

$\longrightarrow$ Stat

$\longrightarrow$ Dist

$\longrightarrow$ BINM

$\longrightarrow$ Bpd — Binomial point distribution (same rules as on page 30)

$\longrightarrow$ Var

Make sure the setting is Var, not List.
Note: This will revert back to List when your calculator is reset.

Number of foot bones.

Number of trials.

Probability.

$$P(X = 3) = 0.0879$$

3 Using tables:

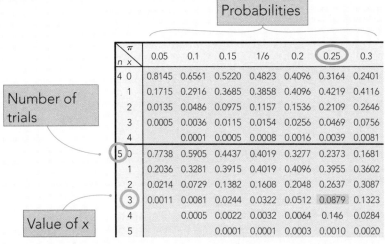

	π						
n x	0.05	0.1	0.15	1/6	0.2	0.25	0.3
4 0	0.8145	0.6561	0.5220	0.4823	0.4096	0.3164	0.2401
1	0.1715	0.2916	0.3685	0.3858	0.4096	0.4219	0.4116
2	0.0135	0.0486	0.0975	0.1157	0.1536	0.2109	0.2646
3	0.0005	0.0036	0.0115	0.0154	0.0256	0.0469	0.0756
4		0.0001	0.0005	0.0008	0.0016	0.0039	0.0081
5 0	0.7738	0.5905	0.4437	0.4019	0.3277	0.2373	0.1681
1	0.2036	0.3281	0.3915	0.4019	0.4096	0.3955	0.3602
2	0.0214	0.0729	0.1382	0.1608	0.2048	0.2637	0.3087
3	0.0011	0.0081	0.0244	0.0322	0.0512	0.0879	0.1323
4		0.0005	0.0022	0.0032	0.0064	0.146	0.0284
5			0.0001	0.0001	0.0003	0.0010	0.0020

Labels: Probabilities; Number of trials; Value of x

$$P(X = 3) = 0.0879$$

Answer in a sentence:

The probability that three out of the five bones are from the feet is 0.0879.

Example 2: Only 10% of the cells in the human body are of human origin. The rest are mostly bacteria and fungi. If 10 cells are selected at random from the human body, what is the probability that more than two are human?

$$p = 0.1$$
$$n = 10$$
$$x = 3, 4, 5, 6, 7, 8, 9, 10$$

> Whatever method you use, it will be easier to calculate $1 - \{P(X = 0) + P(X = 1) + P(X = 2)\}$

1 Using the formula:

$$P(X > 2) = 1 - ((^{10}C_0 (0.1)^0 (0.9)^{10}) + (^{10}C_1 (0.1)^1 (0.9)^9) + (^{10}C_2(0.1)^2(0.9)^8))$$

$$= 1 - (0.3487 + 0.3874 + 0.1937)$$

$$P(X > 2) = 0.0702$$

> The blue brackets are needed if you do the whole calculation on your calculator.

2 Using your calculator:

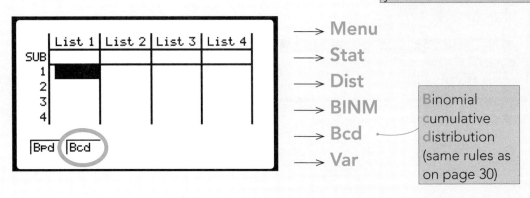

$\longrightarrow$ **Menu**
$\longrightarrow$ **Stat**
$\longrightarrow$ **Dist**
$\longrightarrow$ **BINM**
$\longrightarrow$ **Bcd**
$\longrightarrow$ **Var**

> **B**inomial **c**umulative **d**istribution (same rules as on page 30)

     ISBN: 9780170446938

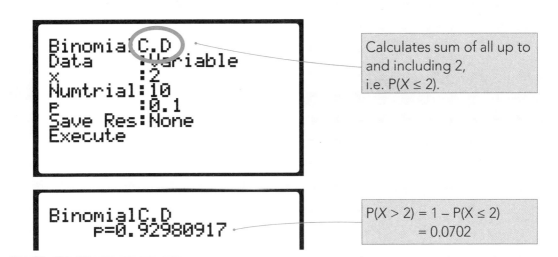

Binomial C.D
Data   :Variable
X   :2
Numtrial:10
P   :0.1
Save Res:None
Execute

Calculates sum of all up to and including 2, i.e. P(X ≤ 2).

BinomialC.D
    p=0.92980917

$P(X > 2) = 1 - P(X \le 2)$
$= 0.0702$

$P(X > 2) = 0.0702$

3 Using tables:

Either:

10	0	0.5987	0.3487	0.1969	0.1615
	1	0.3151	0.3874	0.3474	0.3230
	2	0.0746	0.1937	0.2759	0.2907
	3	0.0105	0.0574	0.1298	0.1550
	4	0.0010	0.0112	0.0401	0.0543
	5	0.0001	0.0015	0.0085	0.0130
	6		0.0001	0.0012	0.022
	7			0.0001	0.0002
	8				
	9				
	10	(all other entries < 0.0001)			

Add these numbers
→ 0.9298

It **is** possible that 7, 8, 9 or 10 human cells were selected, but the probability for each is < 0.0001. However, their combined probabilities could add to > 0.0001 ∴ the listed probabilities could add to 0.999, not 1.

$P(X > 2) = 1 - 0.9298$
$= 0.0702$

Or:

10	0	0.5987	0.3487	0.1969	0.1615
	1	0.3151	0.3874	0.3474	0.3230
	2	0.0746	0.1937	0.2759	0.2907
	3	0.0105	0.0574	0.1298	0.1550
	4	0.0010	0.0112	0.0401	0.0543
	5	0.0001	0.0015	0.0085	0.0130
	6		0.0001	0.0012	0.022
	7			0.0001	0.0002
	8				
	9				
	10	(all other entries < 0.0001)			

Add these numbers
→ 0.0702

$P(X > 2) = 0.0702$

Answer the following questions.

1 Consider a binomial distribution with $n = 6$ and $p = 0.1$.
Complete the table.

	Values **x** can take	Probability
P(x = 0)	0 1 2 3 4 5 6	0.5314
P(x = 2)	0 1 2 3 4 5 6	
P(x = 6)	0 1 2 3 4 5 6	
P(x < 2)	0 1 2 3 4 5 6	
P(x > 1)	0 1 2 3 4 5 6	
P(x ≥ 3)	0 1 2 3 4 5 6	
P(1 < x < 4)	0 1 2 3 4 5 6	
P(x ≤ 3)	0 1 2 3 4 5 6	
P(3 < x ≤ 6)	0 1 2 3 4 5 6	

What do you notice about your last two answers? Explain your answer.

_____

2 Consider a binomial distribution with $n = 10$ and $p = 0.4$.
Complete the table.

	Values **x** can take	Probability
P(x is 0)		0.0060
P(x is between 7 and 9)		
P(x equals 10)		
P(x is at most 2)		
P(x is 4 or more)	4, 5, 6, 7, 8, 9, 10	
P(x is between 3 and 8)		
P(x is at least 6 and at most 10)		
P(x is at least 2)		
P(x is no more than 1)		

Calculate the mean of this distribution.

_____

Calculate the standard deviation of this distribution.

_____

3   The school canteen sells two flavours of Juicy ice blocks: tropical and orange. 45% of those sold are tropical-flavoured Juicies.

a What is the theoretical probability that exactly five out of the next eight Juicies sold are tropical flavoured?

$p =$ _____ $n =$ _____ $x =$ _____

_____

b What is the theoretical probability that fewer than four of the next eight Juicies sold are tropical flavoured?

$p =$ _____ $n =$ _____ $x =$ _____

_____

c What is the theoretical probability that at least six of the next eight Juicies sold are tropical flavoured?

$p =$ _____ $n =$ _____ $x =$ _____

_____

d What is the theoretical probability that exactly five of the next eight Juicies sold are orange flavoured? (Hint: If five are orange flavoured, how many are tropical flavoured?)

$p =$ _____ $n =$ _____ $x =$ _____

_____

e What is the theoretical probability that no fewer than six of the next eight Juicies sold are orange flavoured?

$p =$ _____ $n =$ _____ $x =$ _____

_____

f Write down four reasons why you think that the binomial distribution is appropriate for this situation. Explain your reasons.

_____

_____

_____

_____

_____

_____

_____

_____

_____

4   The probability of throwing a six using a fair die is $\frac{1}{6}$.

   a If a die is thrown 10 times, what is the probability of not getting any sixes?

   $p =$ _____ $n =$ _____ $x =$ _____

   _____

   b If a die is thrown 10 times, what is the probability of getting zero, one or two sixes?

   $p =$ _____ $n =$ _____ $x =$ _____

   _____

   c If a die is thrown 10 times, what is the probability of getting three sixes or more?

   $p =$ _____ $n =$ _____ $x =$ _____

   _____

   d Explain the relationship between your last two answers.

   _____

   _____

   e Calculate the mean and standard deviation for the number of sixes obtained from
      10 throws of a die.

   _____

   f Meriama and Matt each throw a die 10 times. Calculate the probability that both
      throw exactly one six.

   _____

   _____

   g Write down four reasons why you think that the binomial distribution is appropriate
      for this situation. Explain your reasons.

   _____

   _____

   _____

   _____

   _____

   _____

   _____

     ISBN: 9780170446938

5   Amanda and her friends love fruity chews. They come in packets of 12, and the manufacturer claims each fruity chew is equally likely to be orange, raspberry, lemon, lime or passionfruit flavoured.

   a What is the theoretical probability that a packet contains no orange-flavoured chews?

   b What is the theoretical probability that a packet contains fewer than four orange-flavoured chews?

   c Calculate the mean and standard deviation for the number of orange-flavoured chews in each packet.

   d Amanda and Nigel buy a packet of fruity chews each. Calculate the theoretical probability that there is exactly one orange fruity chew in each packet.

   e What is the theoretical probability that a packet contains at least five chews that are either raspberry or passionfruit flavoured?

   f Write down four reasons why you think that the binomial distribution is appropriate for this situation. Explain your reasons.

6 A camping ground owner has kept a record of how many people stay each night in each of his cabins. Each cabin contains six single bunks. The probability that a bunk is occupied is 0.35.

a On any given night, what is the theoretical probability that a cabin has just one bunk occupied?

b On any given night, what is the theoretical probability that a cabin has at least three bunks occupied?

c Calculate the theoretical probability that a cabin is unoccupied for two consecutive nights.

d The graph shows the observed (experimental) distribution for the number of bunks occupied each night for the cabins. The camping ground owner has partly completed a graph comparing this data with the theoretical binomial distribution (——). Complete the graph.

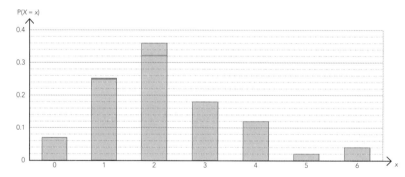

e Discuss what conclusions the camping ground owner could draw from the completed graph.

 ISBN: 9780170446938

7 A teacher gave the 28 students in her class a 'frivolous' quiz. This quiz has six exotic and weird questions to which students could not be expected to know the answer. The questions are multiple choice with four possible answers, of which only one is correct.

The table shows the number of correct answers obtained by the 28 students.

Number correct	0	1	2	3	4	5	6
Number of students	2	9	6	6	4	1	0
Probability	0.0714						
Binomial distribution		0.3560					

Experimental probabilities.

Theoretical binomial probabilities.

a Complete the table to show the actual (experimental) probability distribution for the number of correct answers from the students.

b The teacher claims that a binomial distribution would be suitable to show the expected probabilities if a computer was programmed to randomly select an answer for each question (or if a large number of chimpanzees sat the test). Do you agree? Justify your answer.

_____

_____

_____

_____

_____

c Complete the bottom row of the table to show an appropriate theoretical binomial distribution.

d Is there evidence that the class did better than expected results from the computer (or the chimpanzees)? You should include calculations as well as the evidence from the table in your answer.

_____

_____

_____

_____

_____

_____

8 It is possible to insert a small amount of lead into a die so that it becomes 'weighted'. The probabilities for throwing a 1, 2, 3, 4, 5 or 6 are not all equal in a weighted die. Sam claims he has a die that is weighted so that it is more likely than normal dice to produce a six. Toby doesn't believe him. As evidence, Sam throws the die 24 times and records the number of sixes. The results are shown in the table.

	Outcomes	
	Sixes	Non-sixes
24 throws	6	18

Use the binomial distribution to investigate whether or not Sam really has a weighted die. You should support your answer with statistical reasoning and calculations.

_____

_____

_____

_____

_____

_____

_____

_____

_____

9 Some situations can be modelled by more than one distribution. Every metre of some very expensive fabric is inspected for faults as it comes off the production line. It is found that the probability of one fault is 0.05 for every metre.

a Complete the top row of the table to show the theoretical binomial probability distribution for the number of faults in 10 one-metre-lengths of fabric.

	0	1	2	3	4	5
Binomial			0.0746			
Poisson	0.6065					

b Calculate the mean number of faults in a 10-metre length.

_____

c State an appropriate value for *lambda*, and complete the bottom row of the table to show the theoretical Poisson distribution for the number of faults in a 10-metre length.

_____

d Compare the binomial and Poisson distributions for this situation. Can you explain your findings?

_____

_____

_____

 ISBN: 9780170446938

Inverse binomial problems

- As with inverse Poisson problems, you can do these **only** if you are given $P(X = 0)$.
- You also need to be told either n or p.

$$P(X = 0) = {}^nC_0\, p^0\, (1 - p)^{n-0}$$

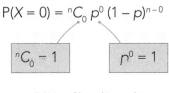

$$\therefore P(X = 0) = (1 - p)^n$$

Given $n \Rightarrow$ find p

Example: A company sells packets of 10 pen drives. Customers return 14.9% of packs because their pack contains at least one faulty flash drive. Calculate the percentage of flash drives that are faulty.

$$n = 10$$
$$P(X = 0) = 1 - 0.149$$
$$= 0.8510$$

$$P(X = 0) = {}^nC_0\, p^0\, (1 - p)^{n-0}$$
$$0.8510 = (1 - p)^{10}$$
$$1 - p = \sqrt[10]{0.8510}$$
$$1 - p = 0.9840$$
$$p = 0.0160$$

$$\therefore 1.6\% \text{ of pen drives are faulty.}$$

Given $p \Rightarrow$ find n

Example: The company would like to reduce the number of packets of flash drives returned to less than 10%. Given that they use the same flash drives, of which 1.6% are faulty, calculate the maximum number that they can put in each packet. Calculate the percentage of new packs which will be returned by customers because they contain faulty flash drives.

$$p = 0.0160$$
$$1 - p = 0.9840$$
$$P(X = 0) = 1 - 0.1$$
$$= 0.9$$

$$P(X = 0) = {}^nC_0\, p^0\, (1 - p)^{n-0}$$
$$0.9 = (0.9840)^n$$
$$n \log 0.9840 = \log 0.9$$
$$n = \frac{\log 0.9}{\log 0.9840}$$
$$n = 6.5322$$

$$\therefore \text{ They should put six flash drives in each packet.}$$

$$n = 6 \rightarrow P(X = 0) = {}^6C_0\, p^0\, (0.9840)^6$$
$$P(X = 0) = 0.9078$$
$$\therefore P(X > 0) = 1 - 0.9078$$
$$= 0.0922$$

$$\therefore \text{ They can expect 9.22\% of packets to be returned.}$$

Answer the following questions.

1 a A department store sells boxes of six dinner plates. 22.7% of the boxes are returned because they contain at least one broken plate. Calculate the percentage of dinner plates that are broken.

$P(X = 0) =$ _____

_____

_____

_____

_____

_____

_____

b The department store manager is not happy. He would like to see the proportion of returned boxes reduced to 13%. Calculate the maximum percentage of broken dinner plates required in order to achieve this.

_____

_____

_____

_____

_____

_____

_____

c Another way to reduce the proportion of returned boxes would be to reduce the number of plates in each box. Using your answer from part **a**, calculate the maximum number that could be in each box in order to achieve a maximum return rate of 13%. Calculate the expected return rate for the new, smaller boxes.

_____

_____

_____

_____

_____

_____

2   Mick grows courgettes for the local market. He puts 20 plants in each row, but some plants don't thrive. Over the years he has found that all 20 plants thrive in an average of 39% of his rows. Calculate the probability that a plant thrives.

3   Amira is tossing a fair coin. How many times must she toss the coin in order to be at least 95% certain of getting at least one head? If she tossed her coin the required number of times, what is the probability that she will get at least one head?

4   A car mechanic gives a small packet of jellybeans to each customer who has their car serviced. Jellybeans come in equal numbers of 10 different colours. The logo for his company is red. He would like a minimum of one red jellybean to be in at least two thirds of the packets. How many jellybeans will he need to put in each packet, and what percentage of packets will have at least one red jellybean?

Continuous random variables

1 The normal distribution

Characteristics of the normal distribution:

- It is a **continuous** distribution, used with **measured data** such as distance, mass, area, etc.
- Most data is **clustered around a central value**, with a few extreme values either side.
- There should be **no upper or lower limit** to the values it can take.
- No upper or lower limit means that the curve **never touches** the x axis.
- The distribution is **symmetrical**.
- It always takes this shape:

When sketching a normal distribution you must leave a small gap between the curve and the x axis.

The parameters for the normal distribution are:

- the **mean** (μ or $\bar{x}$) – this is at the centre of the distribution.
- the **standard deviation** (σ or s) – this measures the spread of the distribution, and occurs at the point of inflection in the curve. Nearly all values lie within three standard deviations of the mean.

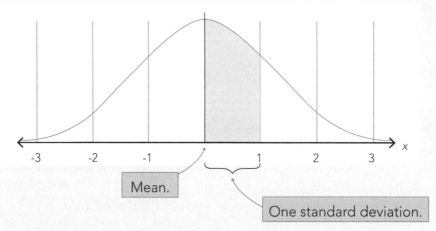

Mean.

One standard deviation.

However, normal distribution curves vary widely, depending on their means and standard deviations.

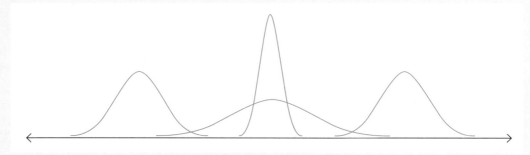

Consequently, we need a **standard normal distribution**.

ISBN: 9780170446938

Sketching normal distributions

- The **mean** will be in the **middle** of the curve.
- 95% of observations lie within ± 2 standard deviations of the mean, so the total **width of the curve should be a little more than 2 standard deviations both sides of the mean**.
- 99% of observations lie within ± 3 standard deviations of the mean, so the total **width of the curve should be about 3 standard deviations both sides of the mean**.
- Remember that the **areas** under normal distribution curves are **equal** (ie, add to 1).
- This means that a curve with a **smaller standard deviation** (narrower) will be **taller** than one with a large standard deviation.
- Be sure to draw your curves so they don't quite touch the x axis.

Example: On the same set of axes, sketch the probability distribution models for the following:

1 A normal distribution with a mean of 8 and a standard deviation of 1.5. _____
2 A normal distribution with a mean of 5 and a standard deviation of 1.0. _____

Step 1: Mark the mean for model 1 (8).

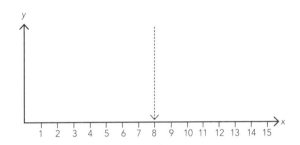

Step 2: Estimate the width for model 1: draw marks at ± 2 standard deviations. In this case: ± 3 units.

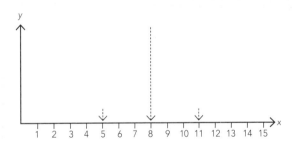

Step 3: Sketch the curve of model 1, extending it slightly beyond ± 3 units.

Be sure that there is a small gap between the x axis and the curve.

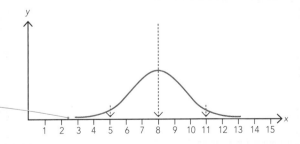

Step 4: Repeat steps 1 to 3 for the curve of model 2, extending it slightly beyond ±
2 units.

Notice that curve 2 **must** be taller
because it is narrower due to the
smaller standard deviation, and the
areas under both curves must be the
same.

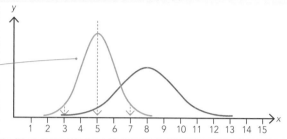

1 On the graph, write the letter which best matches the pairs of parameters for each of
the following probability distribution models:

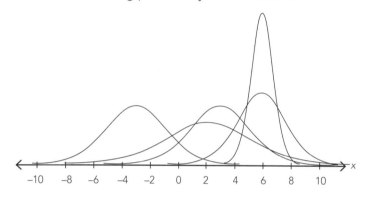

A: Mean: 6
Standard deviation: 0.8

B: Mean: 3
Standard deviation: 2.1

C: Mean: 2.1
Standard deviation: 3

D: Mean: 6
Standard deviation: 1.7

E: Mean: –3
Standard deviation: 2.1

On the same graph, sketch normal probability distribution models for the following pairs of
data:

2

Distribution	A	B
Mean	4	7
Standard deviation	1.2	2

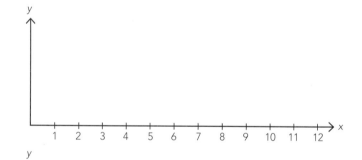

3

Distribution	C	D
Mean	51	59
Standard deviation	13	7

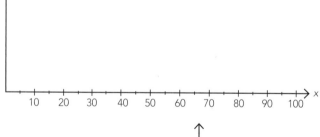

4

Distribution	E	F
Mean	–11	2.2
Standard deviation	2	3.5

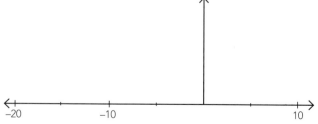

The standard normal distribution

The probability of an event lying between two values is represented by the **area** under the curve. Calculating this is very complicated, so we have tables of probabilities that can be looked up. Because we can't have a set of tables for every different distribution, we convert individual normal distributions to the **standard normal distribution**.

The **parameter** for the **standard normal distribution** is Z: For any value of x, Z measures the number of standard deviations to the right or left of the mean.

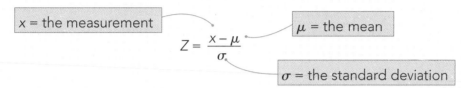

$$Z = \frac{x - \mu}{\sigma}$$

x = the measurement
μ = the mean
σ = the standard deviation

Example 1: Consider a distribution of heights, with a mean of 150 cm and a standard deviation of 5 cm.

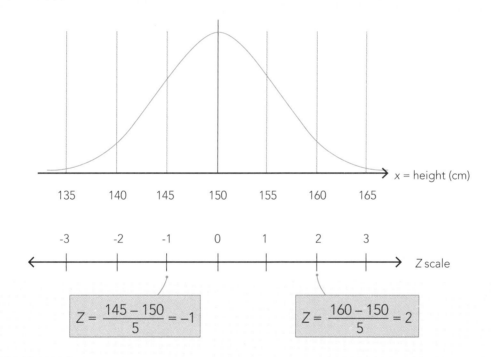

$$Z = \frac{145 - 150}{5} = -1$$

$$Z = \frac{160 - 150}{5} = 2$$

Example 2: The mass of a block of cheese has a mean of 1004 g with a standard deviation of 2 g.
Calculate the Z values for:

a 1007 g: $Z = \dfrac{x - \mu}{\sigma} = \dfrac{1007 - 1004}{2} = 1.5$

b 999 g: $Z = \dfrac{x - \mu}{\sigma} = \dfrac{999 - 1004}{2} = -2.5$

Calculate the value of Z for each of the following situations.

1 The brains of teenagers have a mean weight of 1205 g, with a standard deviation of 21 g. Calculate Z values for teenage brains that weigh:

  a 1247 g

  _____

  b 1142 g

  _____

2 The lengths of jelly snakes have a mean of 255 mm, with a standard deviation of 2 mm. Calculate Z values for jelly snakes with the following lengths:

  a 255 mm

  _____

  b 248 mm

  _____

3 The areas of classrooms in a school have a mean of 56.3 m^2, with a standard deviation of 2.1 m^2. Calculate Z values for classrooms with the following areas:

  a 59.2 m^2

  _____

  b 49.5 m^2

  _____

4 The average volume of blood in an adult is 4.7 L, with a standard deviation of 0.31 L. Calculate Z values for adults with the following blood volumes:

  a 3.9 L

  _____

  b 4.9 L

  _____

5 The average length of time taken by a group of four-year-olds to complete a puzzle was 3 minutes and 22 seconds, with a standard deviation of 16 seconds. Calculate Z values for four-year-olds with the following times. (Convert these times to seconds first.)

  a 3 minutes and 44 seconds

  _____

  b 2 minutes and 42 seconds

  _____

Using Z to calculate a probability

Remember the range of possible values for probabilities:

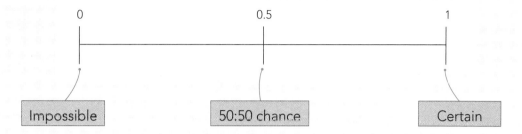

- If something is **certain** to occur, its probability is 1.
- We know that the area under a normal distribution curve represents probabilities of everything that can occur, so that area must be **1**.
- The normal curve is **symmetrical**, so the area under each half must be 0.5.

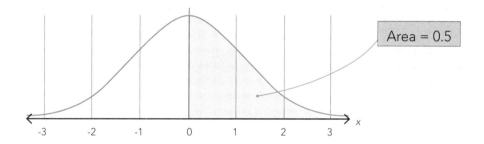

- From the tables we also know these values:

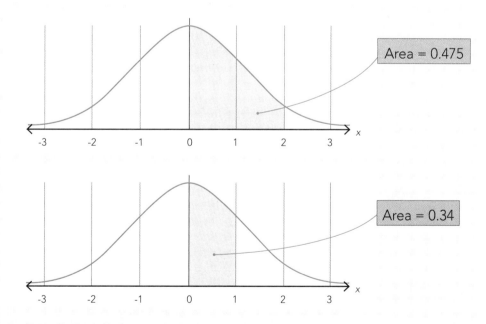

- Tables and graphics calculators can give us probabilities for **any** values of Z.
- Because the graph is **symmetrical**, we can use these to calculate any probabilities under a normal distribution curve.

ISBN: 9780170446938

Reading normal distribution tables

In assessments, you will be given a copy of these to use.

Shows the probability area.

You are given the formula for Z.

Grey areas show probabilities.

Blue areas show Z values.

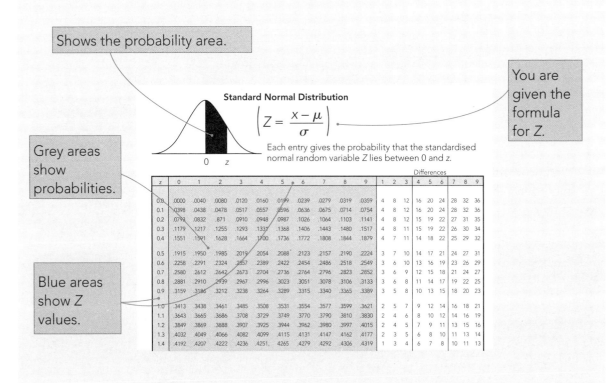

Standard Normal Distribution

$$\left(Z = \frac{x - \mu}{\sigma} \right)$$

Each entry gives the probability that the standardised normal random variable Z lies between 0 and z.

How to use the table:

$$Z = 0.456 \Rightarrow p = 0.1736 + 0.0022 = 0.1758$$

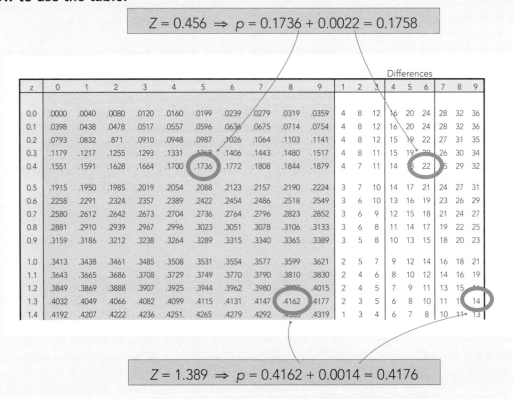

| | | | | | | | | | | | Differences | | | | | | | | |
z	0	1	2	3	4	5	6	7	8	9	1	2	3	4	5	6	7	8	9
0.0	.0000	.0040	.0080	.0120	.0160	.0199	.0239	.0279	.0319	.0359	4	8	12	16	20	24	28	32	36
0.1	.0398	.0438	.0478	.0517	.0557	.0596	.0636	.0675	.0714	.0754	4	8	12	16	20	24	28	32	36
0.2	.0793	.0832	.871	.0910	.0948	.0987	.1026	.1064	.1103	.1141	4	8	12	15	19	22	27	31	35
0.3	.1179	.1217	.1255	.1293	.1331	.1368	.1406	.1443	.1480	.1517	4	8	11	15	19	22	26	30	34
0.4	.1551	.1591	.1628	.1664	.1700	.1736	.1772	.1808	.1844	.1879	4	7	11	14	18	22	25	29	32
0.5	.1915	.1950	.1985	.2019	.2054	.2088	.2123	.2157	.2190	.2224	3	7	10	14	17	21	24	27	31
0.6	.2258	.2291	.2324	.2357	.2389	.2422	.2454	.2486	.2518	.2549	3	6	10	13	16	19	23	26	29
0.7	.2580	.2612	.2642	.2673	.2704	.2736	.2764	.2796	.2823	.2852	3	6	9	12	15	18	21	24	27
0.8	.2881	.2910	.2939	.2967	.2996	.3023	.3051	.3078	.3106	.3133	3	6	8	11	14	17	19	22	25
0.9	.3159	.3186	.3212	.3238	.3264	.3289	.3315	.3340	.3365	.3389	3	5	8	10	13	15	18	20	23
1.0	.3413	.3438	.3461	.3485	.3508	.3531	.3554	.3577	.3599	.3621	2	5	7	9	12	14	16	18	21
1.1	.3643	.3665	.3686	.3708	.3729	.3749	.3770	.3790	.3810	.3830	2	4	6	8	10	12	14	16	19
1.2	.3849	.3869	.3888	.3907	.3925	.3944	.3962	.3980	.3997	.4015	2	4	5	7	9	11	13	15	14
1.3	.4032	.4049	.4066	.4082	.4099	.4115	.4131	.4147	.4162	.4177	2	3	5	6	8	10	11	1	14
1.4	.4192	.4207	.4222	.4236	.4251.	.4265	.4279	.4292	.4306	.4319	1	3	4	6	7	8	10	11	13

$$Z = 1.389 \Rightarrow p = 0.4162 + 0.0014 = 0.4176$$

ISBN: 9780170446938

Given the following Z values, find the probabilities.

1 $Z = 2.1 \Rightarrow P(x) =$

_____

2 $Z = 1.32 \Rightarrow P(x) =$

_____

3 $Z = 0.741 \Rightarrow P(x) =$

_____

4 $Z = 1.035 \Rightarrow P(x) =$

_____

5 $Z = 1.96 \Rightarrow P(x) =$

_____

6 $Z = 3.87 \Rightarrow P(x) =$

_____

Calculating probabilities

Calculating a probability on the right side of the curve

Example: The weights of students' backpacks are found to be normally distributed with a mean of 3.3 kg and a standard deviation of 0.3 kg. Calculate the probability that a backpack weighs between 3.3 kg and 4.0 kg.

Step 1: Draw the picture.

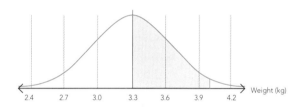

Step 2: Calculate the Z value.

$$Z = \frac{x - \mu}{\sigma} = \frac{4 - 3.3}{0.3} = 2.333$$

Always round Z values to 3 dp.

Step 3: Look up the tables to find the probability.

$$Z = 2.333 \Rightarrow P(3.3 < x < 4.0) = 0.4902$$

Step 4: Write your answer as a sentence and in context.

The probability that a backpack weighs between 3.3 kg and 4.0 kg is 0.4902.

Try for yourself:

Pumpkins from a large crop are found to have a mean weight of 4.1 kg with a standard deviation of 0.45 kg. Calculate the probability that a pumpkin weighs between 4.1 kg and 5 kg.

Step 1:    Step 2:

Step 3:

Step 4:

Probabilities below a value

Remember that the curve is symmetrical, so the area to the left of the mean is 0.5.

Example: The weights of students' backpacks are found to be normally distributed with a mean of 3.3 kg and a standard deviation of 0.3 kg. Calculate the probability that a backpack weighs less than 3.5 kg.

Step 1: Draw the picture.

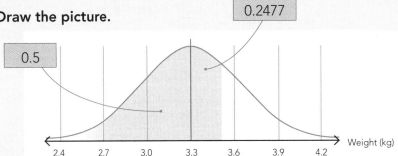

Weight (kg)

Step 2: Calculate the Z value.

$$Z = \frac{x - \mu}{\sigma} = \frac{3.5 - 3.3}{0.3} = 0.667$$

Remember to round to 3 dp.

Step 3: Look up the tables to find the probability.

$$Z = 0.667 \Rightarrow P(3.3 < x < 3.5) = 0.2477$$

$$P(x < 3.5) = 0.5 + 0.2477 = \mathbf{0.7477}$$

Step 4: Write your answer as a sentence and in context.
The probability that a backpack weighs less than 3.5 kg is 0.7477.

Try for yourself:

Pumpkins from a large crop are found to have a mean weight of 4.1 kg with a standard deviation of 0.45 kg. Calculate the probability that a pumpkin weighs less than 5 kg.

Step 1:

Step 2:

Step 3:

Step 4:

Probabilities from the right tail of the curve

Example: The weights of students' backpacks are found to be normally distributed with a mean of 3.3 kg and a standard deviation of 0.3 kg. Calculate the probability that a backpack weighs more than 3.7 kg.

Step 1: Draw the picture.

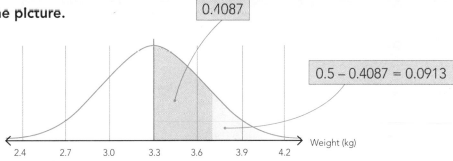

0.1087

0.5 – 0.4087 = 0.0913

Weight (kg)

2.4   2.7   3.0   3.3   3.6   3.9   4.2

Step 2: Calculate the Z value.

$$Z = \frac{x - \mu}{\sigma} = \frac{3.7 - 3.3}{0.3} = 1.333$$

Step 3: Look up the tables to find the probability.

$$Z = 1.333 \implies P(3.3 < x < 3.7) = 0.4087$$

$$P(x > 3.7) = 0.5 - 0.4087 = 0.0913$$

Step 4: Write your answer as a sentence and in context.
The probability that a backpack weighs more than 3.7 kg is 0.0913.

Try for yourself:

Pumpkins from a large crop are found to have a mean weight of 4.1 kg with a standard deviation of 0.45 kg. Calculate the probability that a pumpkin weighs more than 4.7 kg.

Step 1:

Step 2:

Step 3:

Step 4:

Probabilities on the left of the curve

Remember that the curve is symmetrical, so the area to the left is a reflection of that on the right.

Example: The weights of students' backpacks are found to be normally distributed with a mean of 3.3 kg and a standard deviation of 0.3 kg. Calculate the probability that a backpack weighs between 2.7 kg and 3.3 kg.

Step 1: Draw the picture.

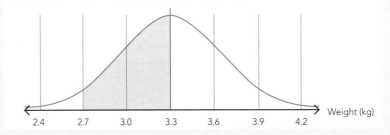

Step 2: Calculate the Z value.

$$Z = \frac{x - \mu}{\sigma} = \frac{2.7 - 3.3}{0.3} = -2$$

Step 3: Look up the tables to find the probability.

$$Z = 2 \Rightarrow P(2.7 < x < 3.3) = 0.4772$$

Step 4: Write your answer as a sentence and in context.

The probability that a backpack weighs between 2.7 kg and 3.3 kg is 0.4772.

Try for yourself:

Pumpkins from a large crop are found to have a mean weight of 4.1 kg with a standard deviation of 0.45 kg. Calculate the probability that a pumpkin weighs between 3 kg and 4.1 kg.

Step 1:

Step 2:

Step 3:

Step 4:

    ISBN: 9780170446938

Probabilities from a tail on the left side of the curve

Remember that the curve is symmetrical, so we can reflect probabilities from the left to the right.

Example: The weights of students' backpacks are found to be normally distributed with a mean of 3.3 kg and a standard deviation of 0.3 kg. Calculate the probability that a backpack weighs less than 2.8 kg.

Step 1: Draw the picture.

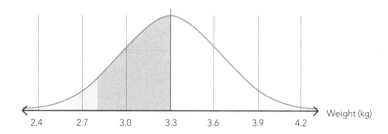

Weight (kg)

Step 2: Calculate the Z value.

$$Z = \frac{x - \mu}{\sigma} = \frac{2.8 - 3.3}{0.3} = -1.667$$

Step 3: Look up the tables to find the probability.

$$Z = 1.667 \implies P(2.8 < x < 3.3) = 0.4523$$

$$P(x < 2.8) = 0.5 - 0.4523 = 0.0477$$

Step 4: Write your answer as a sentence and in context.
The probability that a backpack weighs less than 2.8 kg is 0.0477.

Try for yourself:

Pumpkins from a large crop are found to have a mean weight of 4.1 kg with a standard deviation of 0.45 kg. Calculate the probability that a pumpkin weighs less than 3 kg.

Step 1:

Step 2:

Step 3:

Step 4:

Probabilities from both sides of the curve

Example: The weights of students' backpacks are found to be normally distributed with a mean of 3.3 kg and a standard deviation of 0.3 kg. Calculate the probability that a backpack weighs between 2.9 kg and 3.6 kg.

Step 1: Draw the picture.

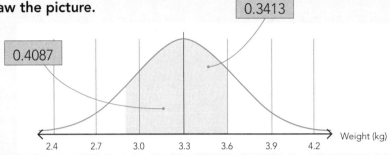

Step 2: Calculate the Z value.

$$Z_{2.9} = \frac{x - \mu}{\sigma} = \frac{2.9 - 3.3}{0.3} = -1.333 \text{ and } Z_{3.6} = \frac{x - \mu}{\sigma} = \frac{3.6 - 3.3}{0.3} = 1$$

Step 3: Look up the tables to find the probability.

$Z = 1.333 \Rightarrow P(2.9 < x < 3.3) = 0.4087$ and $Z = 1 \Rightarrow P(3.3 < x < 3.6) = 0.3413$

$P(2.9 < x < 3.6) = 0.4087 + 0.3413 = 0.75$

The new bit – you need to **add** the two probabilities.

Step 4: Write your answer as a sentence and in context.
The probability that a backpack weighs between than 2.9 kg and 3.6 kg is 0.75.

Try for yourself:

Pumpkins from a large crop are found to have a mean weight of 4.1 kg with a standard deviation of 0.45 kg. Calculate the probability that a pumpkin weighs between 3.5 kg and 4.5 kg.

Step 1:

Step 2:

Step 3:

Step 4:

 ISBN: 9780170446938

Probabilities from two tails of the curve

Remember that the curve is symmetrical, so we can reflect probabilities from the left to the right.

Example: The weights of students' backpacks are found to be normally distributed with a mean of 3.3 kg and a standard deviation of 0.3 kg. Calculate the probability that a backpack weighs less than 2.8 kg or more than 3.6 kg.

Step 1: Draw the picture.

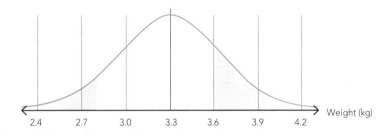

Step 2: Calculate the Z value.

$$Z_{2.8} = \frac{x - \mu}{\sigma} = \frac{2.8 - 3.3}{0.3} = -1.667 \quad \text{and} \quad Z_{3.6} = \frac{x - \mu}{\sigma} = \frac{3.6 - 3.3}{0.3} = 1$$

Step 3: Look up the tables to find the probability.

$Z = 1.667 \Rightarrow P(2.8 < x < 3.3) = 0.4523$ and $Z = 1 \Rightarrow P(3.3 < x < 3.6) = 0.3413$

$$P(x < 2.8 \text{ or } x > 3.6) = (0.5 - 0.4523) + (0.5 - 0.3413) = 0.2064$$

Step 4: Write your answer as a sentence and in context.   **Add** the areas.
The probability that a backpack weighs less than 2.8 kg or more than 3.6 kg is 0.2064.

Try for yourself:

Pumpkins from a large crop are found to have a mean weight of 4.1 kg with a standard deviation of 0.45 kg. Calculate the probability that a pumpkin weighs less than 3.2 kg or more than 4.2 kg.

Step 1:

Step 2:

Step 3:

Step 4:

Calculating probabilities by difference

Example: The weights of students' backpacks are found to be normally distributed with a mean of 3.3 kg and a standard deviation of 0.3 kg. Calculate the probability that a backpack weighs between 2.7 kg and 3.1 kg.

Step 1: Draw the picture.

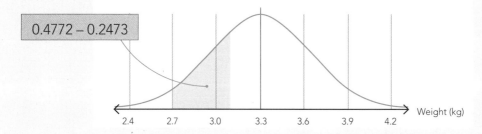

$0.4772 - 0.2473$

Weight (kg)

2.4   2.7   3.0   3.3   3.6   3.9   4.2

Step 2: Calculate the Z value.

$$Z_{2.7} = \frac{x - \mu}{\sigma} = \frac{2.7 - 3.3}{0.3} = -2 \quad \text{and} \quad Z_{3.1} = \frac{x - \mu}{\sigma} = \frac{3.1 - 3.3}{0.3} = -0.6667$$

Step 3: Look up the tables to find the probability.

$Z = 2 \Rightarrow P(2.7 < x < 3.3) = 0.4772$ and $Z = 0.6667 \Rightarrow P(3.1 < x < 3.3) = 0.2477$

$$P(2.7 < x < 3.1) = 0.4772 - 0.2477 = 0.2295$$

Subtract the areas.

Step 4: Write your answer as a sentence and in context.
The probability that a backpack weighs between 2.7 kg and 3.1 kg is 0.2295.

Try for yourself:

Pumpkins from a large crop are found to have a mean weight of 4.1 kg with a standard deviation of 0.45 kg. Calculate the probability that a pumpkin weighs between 3.3 kg and 3.7 kg.

Step 1:

Step 2:

Step 3:

Step 4:

ISBN: 9780170446938

Using a graphics calculator to calculate probabilities

1 Question on page 67:
   The weights of students' backpacks are found to be normally distributed with a mean
   of 3.3 kg and a standard deviation of 0.3 kg. Calculate the probability that a backpack
   weighs between 2.7 kg and 3.1 kg.

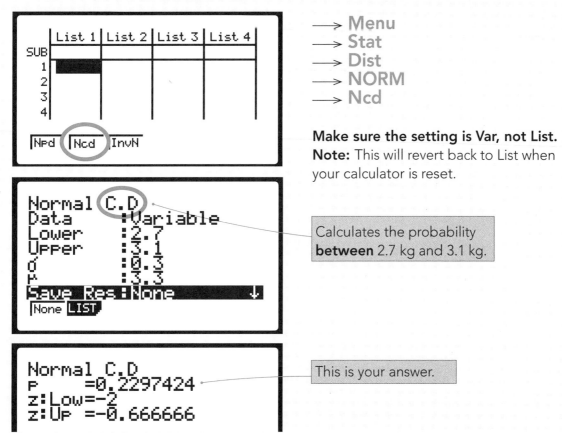

$\longrightarrow$ Menu
$\longrightarrow$ Stat
$\longrightarrow$ Dist
$\longrightarrow$ NORM
$\longrightarrow$ Ncd

Make sure the setting is Var, not List.
Note: This will revert back to List when
your calculator is reset.

Calculates the probability
between 2.7 kg and 3.1 kg.

This is your answer.

2 Question on page 62:
   The weights of students' backpacks are found to be normally distributed with a mean
   of 3.3 kg and a standard deviation of 0.3 kg. Calculate the probability that a backpack
   weighs more than 3.7 kg.

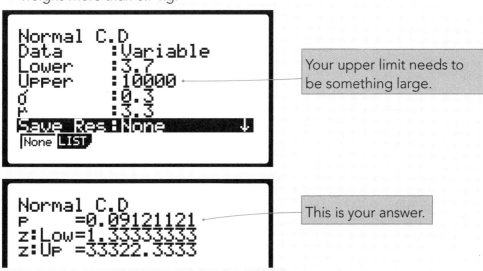

Your upper limit needs to
be something large.

This is your answer.

Mixing it up

1 If the weights of a tuatara (a rare New Zealand ground-living lizard) are normally distributed with a mean of 680 g and a standard deviation of 95 g, answer the following.

a Calculate the percentage of tuatara that weigh less than 700 g.

b Calculate the probability that a tuatara weighs between 600 and 800 g.

c Calculate the probability that a tuatara weighs less than 500 g.

d There are about 50 000 tuatara living on Stephens Island in Cook Strait. About how many of these would you expect to weigh more than 550 g?

e Explain whether or not the normal distribution is appropriate for this situation.

2   The machine winds string into balls, which are advertised as holding lengths of 50 m.
    The machine is set so that the lengths are normally distributed with a mean of 51.50 m
    and a standard deviation of 0.95 m.

    a Calculate the probability that a ball of string contains more than 54 m.

    _____

    _____

    _____

    b Calculate the percentage of balls that have less than the advertised length.

    _____

    _____

    _____

    c If the machine makes 1200 balls of string each day, how many could be expected to
    contain between 51 and 53 m?

    _____

    _____

    _____

    d Two balls are selected at random. What is the probability that both contain less than
    50.5 m each?

    _____

    _____

    _____

    e Explain whether or not the normal distribution is appropriate for this situation.

    _____

    _____

    _____

    _____

    _____

    _____

    _____

    _____

    _____

3 The mean number of pollen grains per gram in samples of honey is found to be normally distributed with a mean of 2170 and a standard deviation of 86.

a Calculate the percentage of samples that have fewer than 2200 pollen grains per gram.

_____

_____

_____

b Calculate the probability that a sample has between 2000 and 2300 pollen grains per gram.

_____

_____

_____

c A researcher takes three samples of honey. Calculate the probability that all three samples have fewer than 2350 pollen grains per gram.

_____

_____

_____

d If 129 samples of honey are taken, how many would be expected to have either fewer than 2050 or more than 2250 pollen grains per gram?

_____

_____

_____

e Give two reasons why the normal distribution may not be appropriate for this situation.

_____

_____

_____

_____

_____

_____

_____

_____

4 The graph shows the probability distribution for marks out of 100 obtained by a large group of students. The standard deviation of these marks is 16.5.

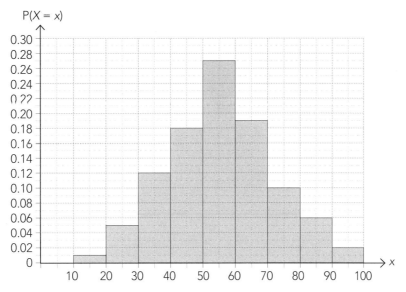

a Calculate the mean mark.

_____

_____

b Use the graph to find the probability that a student earns 70% or more.

_____

c Use the normal distribution to find the probability that a student earns 70% or more.

_____

_____

_____

d Use the same two methods to find the probability that a student earns 40% or less.

_____

_____

e Give reasons for and against applying the normal distribution in this situation.

_____

_____

_____

_____

_____

_____

Give reasons for and against using the normal distributions in the following situations.

5 Body lengths (mm) of a population of mice.

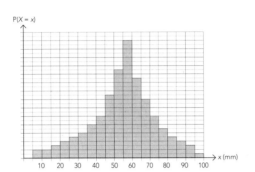

For:

_____

_____

_____

Against:

_____

_____

_____

6 Lengths of rats' tails (mm).

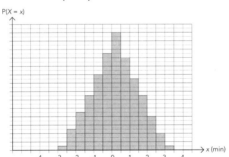

For:

_____

_____

_____

Against:

_____

_____

_____

7 Difference between scheduled time of arrival of a bus and the actual time of arrival (min).

For:

_____

_____

_____

Against:

_____

_____

_____

8 Number of Level 3 NCEA credits obtained by a group of Level 3 students.

For:

_____

_____

_____

Against:

_____

_____

_____

Inverse normal distribution

This is used when:

- we know the probability of an event
- we don't know the value of one of the parameters.
- we use the tables to look up the value of Z which corresponds to the probability
- we use the formula for Z to calculate the missing parameter.

Using a probability to find Z

How to use the tables:

$P(x) = 0.2764 \Rightarrow Z = 0.76$

$P(x) = 0.3991 \Rightarrow Z = 1.276$

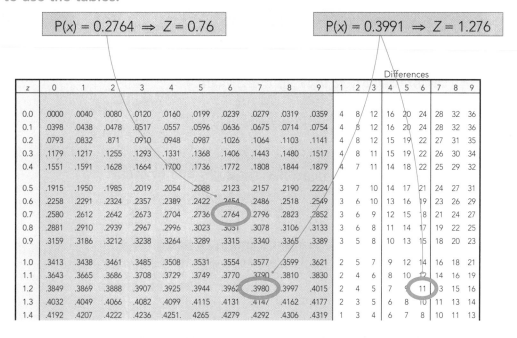

z	0	1	2	3	4	5	6	7	8	9	1	2	3	4	5	6	7	8	9
0.0	.0000	.0040	.0080	.0120	.0160	.0199	.0239	.0279	.0319	.0359	4	8	12	16	20	24	28	32	36
0.1	.0398	.0438	.0478	.0517	.0557	.0596	.0636	.0675	.0714	.0754	4	8	12	16	20	24	28	32	36
0.2	.0793	.0832	.871	.0910	.0948	.0987	.1026	.1064	.1103	.1141	4	8	12	15	19	22	27	31	35
0.3	.1179	.1217	.1255	.1293	.1331	.1368	.1406	.1443	.1480	.1517	4	8	11	15	19	22	26	30	34
0.4	.1551	.1591	.1628	.1664	.1700	.1736	.1772	.1808	.1844	.1879	4	7	11	14	18	22	25	29	32
0.5	.1915	.1950	.1985	.2019	.2054	.2088	.2123	.2157	.2190	.2224	3	7	10	14	17	21	24	27	31
0.6	.2258	.2291	.2324	.2357	.2389	.2422	.2454	.2486	.2518	.2549	3	6	10	13	16	19	23	26	29
0.7	.2580	.2612	.2642	.2673	.2704	.2736	.2764	.2796	.2823	.2852	3	6	9	12	15	18	21	24	27
0.8	.2881	.2910	.2939	.2967	.2996	.3023	.3051	.3078	.3106	.3133	3	6	8	11	14	17	19	22	25
0.9	.3159	.3186	.3212	.3238	.3264	.3289	.3315	.3340	.3365	.3389	3	5	8	10	13	15	18	20	23
1.0	.3413	.3438	.3461	.3485	.3508	.3531	.3554	.3577	.3599	.3621	2	5	7	9	12	14	16	18	21
1.1	.3643	.3665	.3686	.3708	.3729	.3749	.3770	.3790	.3810	.3830	2	4	6	8	10	12	14	16	19
1.2	.3849	.3869	.3888	.3907	.3925	.3944	.3962	.3980	.3997	.4015	2	4	5	7	9	11	13	15	16
1.3	.4032	.4049	.4066	.4082	.4099	.4115	.4131	.4147	.4162	.4177	2	3	5	6	8	10	11	13	14
1.4	.4192	.4207	.4222	.4236	.4251.	4265	.4279	.4292	.4306	.4319	1	3	4	6	7	8	10	11	13

Look up the values of Z for each of the following probabilities.

1 $P(x) = 0.4956 \Rightarrow Z = $ _____

2 $P(x) = 0.0478 \Rightarrow Z = $ _____

3 $P(x) = 0.4989 \Rightarrow Z = $ _____

4 $P(x) = 0.4034 \Rightarrow Z = $ _____

5 $P(x) = 0.4508 \Rightarrow Z = $ _____

6 $P(x) = 0.2800 \Rightarrow Z = $ _____

7 $P(x) = 0.4859 \Rightarrow Z = $ _____

8 $P(x) = 0.1436 \Rightarrow Z = $ _____

9 $P(x) = 0.4763 \Rightarrow Z = $ _____

10 $P(x) = 0.4945 \Rightarrow Z = $ _____

Using Z to calculate the value of x — where Z is positive

Example: The weights of students' backpacks are found to be normally distributed with a mean of 3.3 kg and a standard deviation of 0.4 kg. Above what weight (w) do 20% of backpacks lie?

Step 1: Draw the picture.

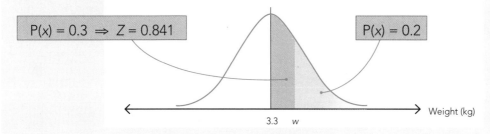

$P(x) = 0.3 \Rightarrow Z = 0.841$

$P(x) = 0.2$

3.3   w

Weight (kg)

Step 2: Use the probability to look up the Z value.

$P(x > w) = 0.2 \Rightarrow P(3.3 < x < w) = 0.5 - 0.2 = 0.3$
So $Z = 0.841$

Step 3: Use the Z value to calculate the value of x.

$Z = \dfrac{w - 3.3}{0.4} = 0.841 \Rightarrow w - 3.3 = 0.3364$, so $w = 3.6364$ kg

Step 4: Write your answer as a sentence and in context.
Twenty per cent of backpacks weigh more than 3.6364 kg.

Step 5: Does my answer make sense?
Yes, because 3.6364 is above the mean of 3.3 kg.

Try for yourself:

The weights of pumpkins from a large crop are found to be normally distributed with a mean weight of 4.1 kg with a standard deviation of 0.45 kg. The heaviest 30% of pumpkins are sold at a higher price. What is the minimum weight for a pumpkin to earn the higher price?

Step 1:

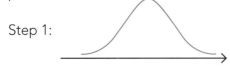

Step 2:

Step 3:

Step 4:

Step 5:

Using *Z* to calculate the value of *x* – where *Z* is negative

Example: The weights of students' backpacks are found to be normally distributed with a mean of 3.3 kg and a standard deviation of 0.4 kg. Below what weight (*w*) do 10% of backpacks lie?

Step 1: Draw the picture.

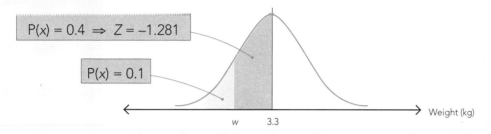

$P(x) = 0.4 \Rightarrow Z = -1.281$

$P(x) = 0.1$

Weight (kg)

w 3.3

Step 2: Use the probability to look up the *Z* value.

$P(x < w) = 0.1 \Rightarrow P(w < x < 3.3) = 0.5 - 0.1 = 0.4$
So $Z = -1.281$ (Note: You could also use $Z = -1.282$)

Step 3: Use the *Z* value to calculate the value of *x*.

$$Z = \frac{w - 3.3}{0.4} = -1.281 \Rightarrow w - 3.3 = -0.5124, \text{ so } w = 2.7876 \text{ kg}$$

Step 4: Write your answer as a sentence and in context.
Ten per cent of backpacks weigh less than 2.787 kg.

Step 5: Does my answer make sense?
Yes, because 2.787 is below the mean of 3.3 kg.

Try for yourself:

The weights of pumpkins from a large crop are found to be normally distributed with a mean weight of 4.1 kg with a standard deviation of 0.45 kg. The smallest 10% of pumpkins are given to a local charity. Calculate the maximum weight for a pumpkin to be given to the charity.

Step 1:

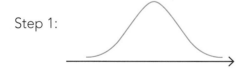

Step 2:

Step 3:

Step 4:

Step 5:

Using a graphics calculator to calculate inverse probabilities

This can be done ONLY for inverse probabilities where you are asked for the value of x, NOT where you need to find the mean or standard deviation.

1 Question on page 75:

The weights of students' backpacks are found to be normally distributed with a mean of 3.3 kg and a standard deviation of 0.4 kg. Above what weight do 20% of backpacks lie?

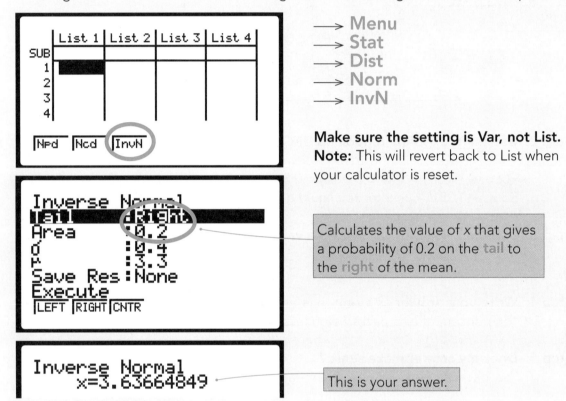

⟶ Menu
⟶ Stat
⟶ Dist
⟶ Norm
⟶ InvN

Make sure the setting is Var, not List.
Note: This will revert back to List when your calculator is reset.

Calculates the value of x that gives a probability of 0.2 on the tail to the right of the mean.

This is your answer.

2 Question on page 76:

The weights of students' backpacks are found to be normally distributed with a mean of 3.3 kg and a standard deviation of 0.4 kg. Below what weight do 10% of backpacks lie?

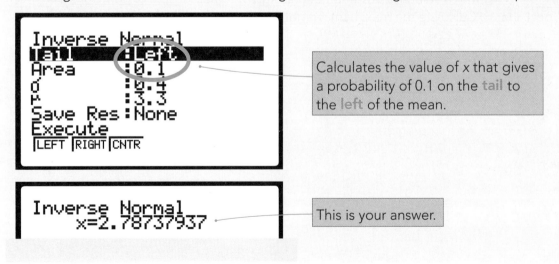

Calculates the value of x that gives a probability of 0.1 on the tail to the left of the mean.

This is your answer.

Using *Z* to calculate the mean

Example: The weights of students' backpacks are found to be normally distributed with a standard deviation of 0.6 kg. If the probability that a backpack weighs more than 4.0 kg is 0.0179, calculate the mean weight of the backpacks.

Step 1: Draw the picture.

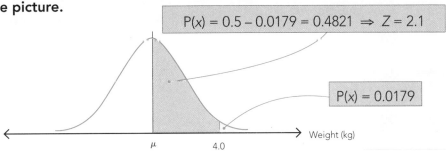

$P(x) = 0.5 - 0.0179 = 0.4821 \Rightarrow Z = 2.1$

$P(x) = 0.0179$

Weight (kg)

μ 4.0

Step 2: Use the probability to look up the *Z* value.

$P(x > 4) = 0.0179 \Rightarrow P(\mu < x < 4) = 0.4821$
From the tables, $Z = 2.1$.

Remember, tables give probabilities only for the area between μ and x:

μ x

Step 3: Use the *Z* value to calculate the value of x.

$$Z = \frac{4 - \mu}{0.6} = 2.1 \Rightarrow 4 - \mu = 1.26, \text{ so } \mu = 2.74 \text{ kg}$$

Step 4: Write your answer as a sentence and in context.
The mean weight for the backpacks is 2.74 kg.

Step 5: Does my answer make sense?
Yes, because 2.74 kg is less than 4 kg.

Try for yourself:

The weights of pumpkins from another large crop are found to be normally distributed with a standard deviation of 0.40 kg. If 5% of these pumpkins weigh more than 4.988 kg, calculate the mean weight of this crop.

Step 1:

Step 2:

Step 3:

Step 4:

Step 5:

Using Z to calculate the standard deviation

Example: The weights of students' backpacks are found to be normally distributed with a mean of 3.3 kg. Calculate the standard deviation if the probability that a backpack weighs between 2.7 kg and 3.3 kg is 0.4332.

Step 1: Draw the picture.

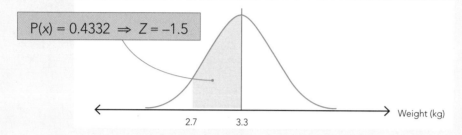

$P(x) = 0.4332 \Rightarrow Z = -1.5$

2.7   3.3   Weight (kg)

Step 2: Use the probability to look up the Z value.

$$P(2.7 < x < 3.3) = 0.4332 \Rightarrow \text{(from the tables) } Z = -1.5$$

Step 3: Use the Z value to calculate the value of x.

$$Z = \frac{2.7 - 3.3}{\sigma} = -1.5 \Rightarrow \frac{-0.6}{\sigma} = -1.5, \text{ so } \sigma = 0.4 \text{ kg}$$

Step 4: Write your answer as a sentence and in context.
The standard deviation for the backpack weights is 0.4 kg.

Step 5: Does my answer make sense?
Yes, because 0.4 x 1.5 = 0.6 kg, which is 3.3 – 2.7 kg.

Try for yourself:

The weights of pumpkins from another large crop are found to be normally distributed with a mean weight of 4.1 kg. If the upper quartile is 4.4 kg, calculate the standard deviation.

Step 1:

Step 2:

Step 3:

Step 4:

Step 5:

Mixing it up

Answer the following.

1 a 15% of the adult male North Island brown kiwi weigh at least 2.3968 kg. If the standard deviation is 0.19 kg and their weights are normally distributed, calculate their mean weight.

b If mean weight of female North Island brown kiwi is 2.8 kg, and 20% weigh less than 2.615 kg, calculate the standard deviation for their weights.

c If the weights of adult male great spotted kiwi are normally distributed with a mean of 2.4 kg and a standard deviation of 0.23 kg, calculate the upper and lower quartiles for the weights of male kiwi.

d If the interquartile range for adult female great spotted kiwi is 0.378 kg, calculate the standard deviation.

e If 6.68% of adult female great spotted kiwi weigh less than 2.88 kg, use your answer from part **d**), to help you calculate their mean weight.

2 a The mean distance run by a player during a top-level, five-set tennis match is 4.8 km. If the distances are normally distributed and the 5% of top players run at least 6.03 km, calculate the standard deviation for the distance run.

b The mean distance run by a player during a top-level basketball game is normally distributed with a mean of 4.1 km and a standard deviation of 0.55 km. Calculate the maximum distance run by a player in the bottom 15% of the distribution.

c In a top-level soccer game, 70% of players run more than 10.1 km. If the distances run by players is normally distributed with a standard deviation of 2.1 km, calculate the mean distance run by each player.

d The mean distance run by a player during a top-level field hockey game is normally distributed with a mean of 9.0 km and a standard deviation of 1.19 km. Calculate the minimum distance run by those in the top 20% of the distribution.

e The distances run during a game by top-level rugby players are normally distributed. The data on those in the central two thirds of this distribution was examined. It was found that the difference between the distances run by those who ran the farthest and those who ran the shortest distances was 2.8 km. Calculate the standard deviation for the distances run.

Mixed normal distribution problems

1 An ornithologist is studying takahe, a New Zealand species of bird that nests on the ground. The weights of adult takahe are known to be approximately normally distributed.

a Adult male takahe are known to have a mean weight of 2.8 kg with a standard deviation of 0.25 kg. Find the probability that a randomly chosen adult male takahe weighs less than 2.5 kg.

b What percentage of adult male takahe could be expected to weigh between 2.8 and 3 kg?

c Adult male takahe that weigh less than 2.5 kg are classified as underweight and are therefore at risk. If there are 134 adult males, how many are at risk?

d What weight is exceeded by 60% of male takahe?

e Adult female takahe have a mean weight of 2.6 kg. It has been found that 10% of the population weigh more than 2.985 kg. Calculate the standard deviation for adult females.

f In an isolated reserve where efforts are being made to breed takahe, the females are fed extra food. While their standard deviation is the same (answer to **e**), their mean weight is larger. 70% of these birds are found to exceed 2.73 kg. Calculate the mean weight for this population of females.

2 Research has shown that the weights of newborn lambs in Southland are normally distributed with a mean of 1.5 kg and a standard deviation of 0.125 kg. Use this model to answer the questions below.

a A newborn lamb in Southland is selected at random. What is the probability that it weighs between 1.5 and 1.7 kg?

b What percentage of newborn lambs in Southland would be expected to weigh more than 1.75 kg?

c Newborn lambs that weigh less than 1.25 kg are underweight and likely to die. If 6400 lambs are born on a farm, how many are likely to be underweight?

d What birth weight is exceeded by 30% of newborn lambs?

e What is the range of the central 70% of weights of newborn lambs?

f In Canterbury the average newborn lamb weight and standard deviation are different. The average weight is 1.7 kg, and it is found that 1.62% of newborn lambs exceed 2 kg. Calculate the standard deviation of lambs in Canterbury.

Continuity correction

Consider the following distributions, remembering that probability is found by finding the area under the curve.

1 A continuous distribution: The distribution of distances (from the attachment point) at which a 10 metre tow rope is expected to break. Consider the probability that the rope will break between any two points.

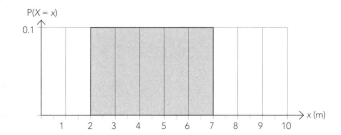

Using the 'area under the curve':

$P(2 \leq X \leq 7) = 0.5$

2 A discrete distribution: The expected distribution of numbers obtained when a 10-sided die is thrown. Consider the probability that the number thrown is between 2 and 7.

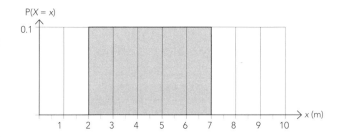

Using the 'area under the curve':

$P(2 \leq X \leq 7) = 0.5$

BUT $P(x = 2, 3, 4, 5, 6 \text{ or } 7) = \mathbf{0.6}$

In order to correct this, we need to consider the area between 1.5 and 7.5.

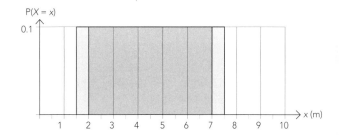

- This is called **continuity correction**.
- It is particularly significant if the **range** of the data is **small**.

It needs to be used when the normal distribution is used for:
1 discrete data
2 continuous data which is rounded, e.g. age in years
3 data from Poisson or binomial distributions, which are always discrete.

THINK before you apply this.
If the range is **small**, then it **WILL make a difference**.
If the range is **large** (e.g. 100), then it will make **very little difference**.
Note: The range $\cong 6 \times$ standard deviation.

Example: If the number of eggs laid by flatback turtles is normally distributed with a mean of 50 and a standard deviation of 7.5, calculate the probability that a turtle lays the following.

a Between 45 and 55 eggs inclusive.

Between 45 and 55 eggs inclusive ⇒ 45, 46, …, 54 or 55 eggs. Applying continuity correction, consider the probability that a turtle lays between **44.5** and **55.5** eggs:

Include an extra 0.5 at each end.

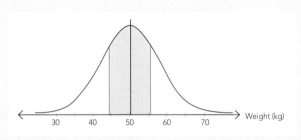

$$Z = \frac{55.5 - 50}{7.5}$$
$$= 0.73$$

$$\therefore P(44.5 \leq X \leq 55.5) = 2 \times 0.2682$$
$$= 0.5364$$

b More than 60 eggs.

Include an extra 0.5 at the left end.

More than 60 eggs ⇒ 61 eggs or more.
Applying continuity correction, consider the probability that a turtle lays **60.5** eggs or more.

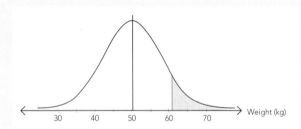

$$Z = \frac{60.5 - 50}{7.5}$$
$$= 1.4$$

$$\therefore P(X \geq 60.5) = 0.5 - 0.4192$$
$$= 0.0808$$

c No more than 45 eggs.

Include an extra 0.5 at the right end.

No more than 45 ⇒ …, 41, 42, 43, 44 or 45.
Applying continuity correction, consider the probability that a turtle lays **45.5** or fewer or 61.5 or more eggs:

$$Z = \frac{45.5 - 50}{7.5}$$
$$= -0.6$$

$$\therefore P(X \leq 45.5) = 0.5 - 0.2258$$
$$= 0.2742$$

Always ADD an extra 0.5 to each end of the area required
(unless the end is a tail).

 ISBN: 9780170446938

Complete the table for a normal distribution (using continunity correction) with a mean of 4 and a standard deviation of 1.0.

Words	Highlight the values that X can take:	Sketch of area	X value(s)	Probability
P(X is greater than 5)	0 1 2 3 4 5 **6 7 8**		5.5	0.0668
P(X is at least 5)	0 1 2 3 4 5 6 7 8			
P(X is less than 4)	0 1 2 3 4 5 6 7 8			
P(X is between 4 and 7)	0 1 2 3 4 5 6 7 8			
P(X is 5)	0 1 2 3 4 5 6 7 8			
P(X is 6 or less)	0 1 2 3 4 5 6 7 8			
P(X is over 2)	0 1 2 3 4 5 6 7 8			
P(X is between 2 and 5 inclusive)	0 1 2 3 4 5 6 7 8			

Complete the table.

Situation	Continuity correction √ or X	Sketch of area	X value(s)	Probability
A test is marked out of 50. The marks are normally distributed with a mean of 31.4 and a standard deviation of 5.2. Calculate the percentage of marks that lie between 25 and 40.	√ (Discrete data)		24.5 and 40.5	86.77%
A group of teenagers have their ages recorded to the nearest year. The ages are normally distributed with a mean of 16.2 years and a standard deviation of 1.1 years. Calculate the probability that the recorded age of a teenager is more than 16.				
A group of adults have their ages recorded to the nearest year. The ages are normally distributed with a mean of 56.8 years and a standard deviation of 12.1 years. Calculate the probability that the recorded age of an adult is less than 45.				
The times taken for the group of adults to complete a task on a website are normally distributed with a mean of 26.8 minutes and a standard deviation of 5.4 minutes. Calculate the probability that an adult completed the task in less than 30 minutes.				
Beck has a 48% success rate when she attempts to shoot netball goals from the edge of the circle. If she has 32 attempts, use the normal distribution to estimate the probability that she is successful in at least 14 of her attempts. (Hint: For binomial, $\mu = np$ and $\sigma = \sqrt{np(1-p)}$)				
The number of items in the pencil cases of a group of students is normally distributed with a mean of 9.4 and a standard deviation of 2.6. Calculate the probability that a student has fewer than six or more than 12 items in their pencil case.				

 ISBN: 9780170446938

2   The rectangular (continuous uniform) distribution

Characteristics of the rectangular distribution:

- It is the **continous** distribution.
- It takes the shape of a **rectangle** because the probabilities are the **same (uniform) for all intervals**.
- It is used where **maximum** and **minimum** are given, but **no mode** is given.

For example, a tug-of-war rope is equally likely to break at any point between the opposing teams.

Parameters for the rectangular distribution:

a = the lowest value the distribution can take
b = the highest value the distribution can take

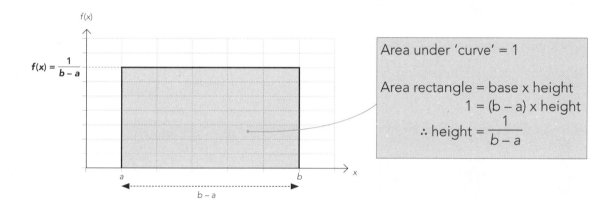

Area under 'curve' = 1

Area rectangle = base x height
$$1 = (b - a) \times \text{height}$$
$$\therefore \text{height} = \frac{1}{b-a}$$

Formula for calculating P($c < x < d$)

$$P(c < x < d) = \frac{d-c}{b-a}$$

This is **not** on your formula sheet.

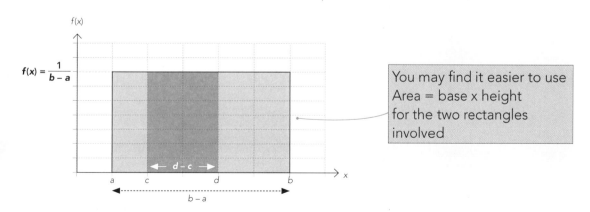

You may find it easier to use
Area = base x height
for the two rectangles
involved

Mean for the rectangular distribution:

$$\text{Mean} = \frac{b+a}{2}$$

Example: Company A requires that all employees work for 40 hours per week, of which a minimum of 10 hours must be spent in the company office. The remaining time can be spent working from home.

a Calculate the probability that an employee spends less than 15 hours in the company office.

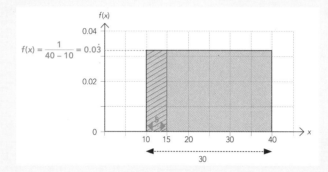

$$P(10 < x < 15) = \frac{5}{30} = 0.1\dot{6}$$

b The 20% of employees who spend the most time working at home do not qualify for a desk of their own. What is the maximum length of time that an employee can work at home each week if they wish to qualify for their own desk?

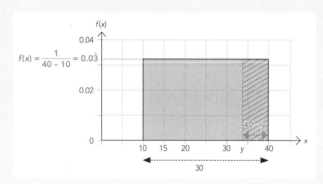

$$P(y < x < 40) = 0.2 = \frac{40 - y}{30}$$
$$40 - y = 0.2 \times 30$$
$$y = 34$$

∴ Maximum time spent working at home is 34 hours.

c Company B has a similar policy, but employees are expected to work 45 hours per week, of which 5 must be in the company office. Sketch the probability distribution for this company on the graph below.

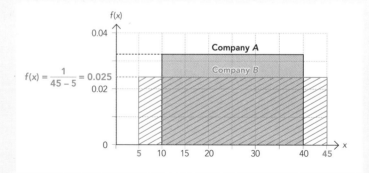

Answer the following.

1 A rectangular distribution has the parameters $a = 4$ and $b = 12$.

    a Show that the height of the distribution $f(x) = 0.125$.

    $a =$ _____ $b =$ _____

    _____

    b Sketch the probability distribution on the axes below. Label the scales and show the values of any significant points.

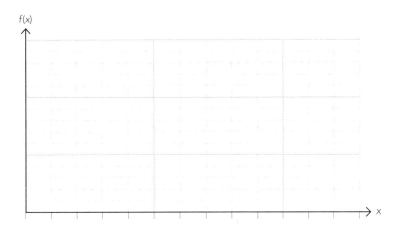

    c Show that the probability that $6 < x < 9 = 0.375$.

    $a =$ _____ $b =$ _____

    _____

2 A rectangular distribution is shown on the graph below.

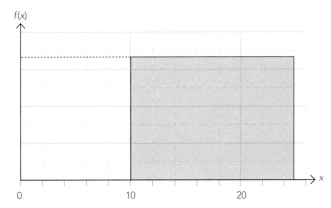

    a Write down the parameters for this distribution.

    _____

    b Calculate the height of the distribution, $f(x)$, and label the vertical axis appropriately.

    _____

    c Calculate the probability that $11 < x < 14$.

    _____

3 A waterski tow rope is 25 metres long. It is equally likely to break at any point.

    a Sketch the probability distribution on the axes below. Label the scales and show the values of any significant points.

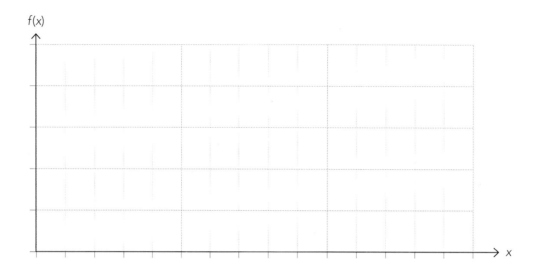

    b Calculate the probability that the rope will break within 4 metres of the skier.

_____

_____

    c Calculate the probability that the rope will break within its central 3 metres.

_____

_____

    d Some similar ski ropes are each 30 metres long. Sketch the probability distribution for these ropes on the axes above. Show the values of any significant points.

    e Calculate the probability that the 30 m ropes break within 2 m of either end.

_____

_____

    f Calculate the mean point at which the 30 m ropes break.

_____

_____

4 Shuttle buses between the domestic and international terminals at Auckland airport leave at 20-minute intervals. The distribution of waiting times is therefore uniform, between 0 and 20 minutes.

a Sketch the probability distribution on the axes below. Label the scales and show the values of any significant points.

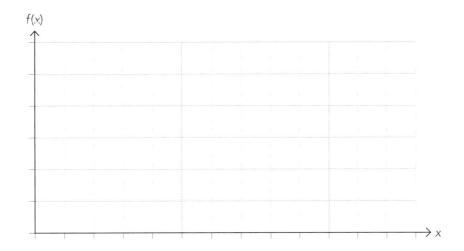

b Calculate the probability that a traveller will have to wait between 2 and 8 minutes.

c Calculate the probability that a traveller will have to wait less than two and a half minutes.

d Two passengers arrive independently to wait for the bus. Calculate the probability that both have to wait for more than 15 minutes.

e If the CEO wanted to change the frequency of the shuttles so that only one traveller in every six had to wait for more than 15 minutes, how often would the shuttles need to run?

f Calculate the value of the y-intercept for this new distribution.

5 A researcher follows the same track each night, looking for a radio-tagged kiwi. The track starts from his hut and finishes at a car park. He knows that he will find it at some point along the track, but not within 100 m of the hut or within 200 m of the car park. He is equally likely to spot the kiwi at any point along the 2.8 km track.

a Draw the probability distribution on the axes below. Label the scales on both axes and show the values of any significant points.

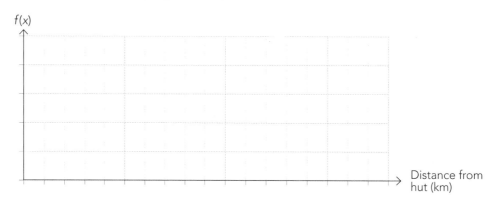

b Calculate the probability that the researcher will find the kiwi within 200 m of the hut.

c Calculate the mean distance from the hut at which he can expect to see the kiwi.

d There is a second radio-tagged kiwi, which lives near a track that leads in the opposite direction from the hut, and ends at a lake. This track is 3.3 km long, and the researcher knows that, like the other, he won't find this kiwi within the first 100 m from the hut. However, he is equally likely to find it at any other point throughout its length. Add this distribution to the graph above and show the values of any significant points.

e Calculate the probability that he will find this kiwi either in the first or the last 500 m of the track.

f One night he walks both tracks, starting each walk from the hut. Calculate the probability that on both tracks he finds both kiwi within the last 500 m of each track.

6 A city council records the times at which it receives 100 phone calls throughout the working day. Their office opens at 8.00 a.m. and closes at 6.00 p.m. They record the total number of calls in each one-hour interval. The probability distribution of their results is shown below.

a Label the *x*-axis with a more appropriate scale.

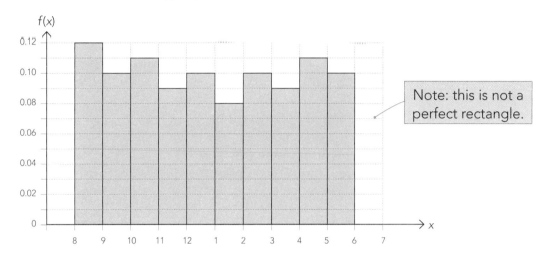

Note: this is not a perfect rectangle.

b A rectangular distribution could be used to model this situation. Write down appropriate parameters for this.

_____

_____

c Using the data in the graph, estimate the proportion of phone calls that are received between 11 a.m. and 4 p.m.

_____

_____

d Use your rectangular model to estimate the proportion of phone calls that are received between 11 a.m. and 4 p.m.

_____

_____

e Identify any limitations to your model or assumptions you have made about the data fit of your model. You may wish to refer to your answers to **c** and **d**.

_____

_____

_____

3 The triangular distribution

Characteristics of the triangular distribution:
- It is a **continuous** distribution.
- It takes the shape of a **triangle**.
- It is used where a **maximum**, a **minimum** and a **mode** are given.

For example, the length of time for a bus trip is between 22 and 27 minutes, with a most likely time of 25 minutes.

Parameters for the triangular distribution:
a = the lowest value the distribution can take
b = the highest value the distribution can take
c = the value of the distribution at its peak

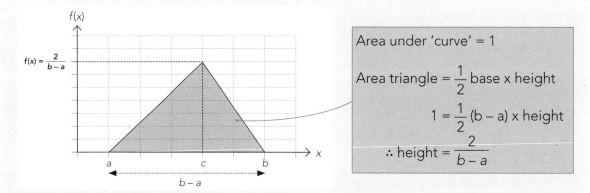

$$f(x) = \frac{2}{b-a}$$

Area under 'curve' = 1

Area triangle $= \dfrac{1}{2}$ base × height

$$1 = \frac{1}{2}(b-a) \times \text{height}$$

$$\therefore \text{height} = \frac{2}{b-a}$$

Formulae for calculating heights at *x* (these are on your formula sheet):

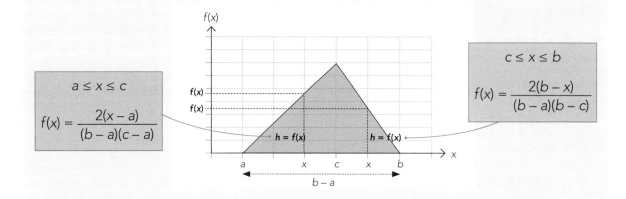

$a \le x \le c$

$$f(x) = \frac{2(x-a)}{(b-a)(c-a)}$$

$c \le x \le b$

$$f(x) = \frac{2(b-x)}{(b-a)(b-c)}$$

Formulae calculating areas:

Area of a triangle $= \dfrac{1}{2} \times$ base × height

Area of a trapezium = average of parallel sides × height

$$= \frac{\text{sum of parallel sides}}{2} \times \text{vertical height}$$

Example: The length of time for a bus trip is between 22 and 27 minutes, with a most likely time of 25 minutes.

a Draw the distribution. Label the scales and the show the values of any intercepts.

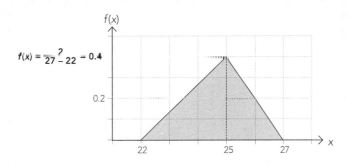

$$f(x) = \frac{2}{27 - 22} = 0.4$$

b Calculate the probability that the bus trip lasts between 26 and 27 minutes.

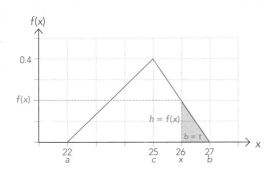

$$\text{height} = f(x) = \frac{2(b - x)}{(b - a)(b - c)}$$

$$= \frac{2(27 - 26)}{(27 - 22)(27 - 25)}$$

$$= \frac{2}{5 \times 2}$$

$$= 0.2$$

$$\text{Probability} = \frac{1}{2} \times \text{base} \times \text{height}$$

$$= \frac{1}{2} \times 1 \times 0.2$$

$$= 0.1$$

c Calculate the probability that the bus trip lies between 23 and 24 minutes.

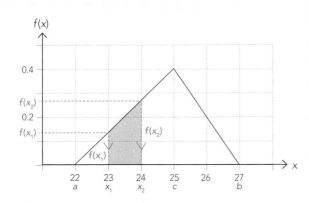

$$f(x_1) = \frac{2(x_1 - a)}{(b - a)(c - a)}$$

$$= \frac{2(23 - 22)}{(27 - 22)(25 - 22)}$$

$$= \frac{2}{5 \times 3}$$

$$= 0.1\dot{3}$$

$$f(x_2) = \frac{2(x_2 - a)}{(b - a)(c - a)}$$

$$= \frac{2(24 - 22)}{(27 - 22)(25 - 22)}$$

$$= \frac{4}{5 \times 3}$$

$$= 0.2\dot{6}$$

$$\text{Area of trapezium} = \frac{\text{sum of parallel sides}}{2} \times \text{vertical height}$$

$$= \frac{0.1\dot{3} + 0.2\dot{6}}{2} \times 1$$

$$= 0.2$$

Answer the following.

1 The graph shows a triangular distribution.

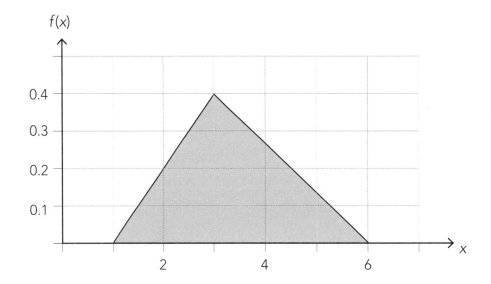

a Write down the parameters for this distribution.

$a =$ _____ $b =$ _____ $c =$ _____

b The maximum height for this distribution is $f(3) = 0.4$. Show how this value was calculated.

_____

_____

c In order to calculate the probability that x is more than five, you would need to know that the triangle height, $f(5)$, is $0.1\dot{3}$. Show how this was calculated.

_____

_____

d Draw lines to show $f(5) = 0.1\dot{3}$ on the graph above.

e Show that the probability that x is more than five is $0.0\dot{6}$.

_____

_____

f Show that $f(2.5) = 0.3$.

_____

_____

g Show that the probability that x lies between 2.5 and 3 is 0.175.

$a =$ _____ $b =$ _____ $c =$ _____ $x_1 =$ _____ $x_2 =$ _____

_____

2 The parameters for a triangular distribution are $a = 4$, $b = 24$ and $c = 16$.

    a Calculate the height, $f(16)$, for this distribution.

    _____

    _____

    b Draw the distribution on the graph below.

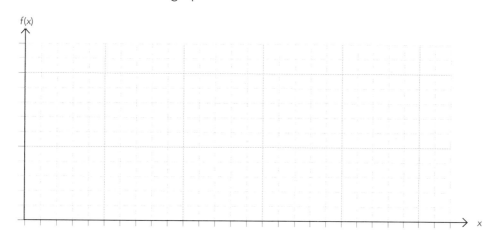

    c Calculate $f(10)$.

    $x =$ _____ $h =$ _____

    _____

    d Calculate the probability that $x < 10$.

    _____

    _____

    e Calculate $f(18)$.

    _____

    _____

    f Calculate the probability that x is between 16 and 18.

    _____

    _____

    g Calculate $P(x > 12)$.

    _____

    _____

    _____

    _____

3 The time taken to replace the battery in a watch varies between 2 and 10 minutes, with the most likely time being 8 minutes.

a Draw the probability distribution on the axes below. Label the scales on both axes and show the values of any significant points.

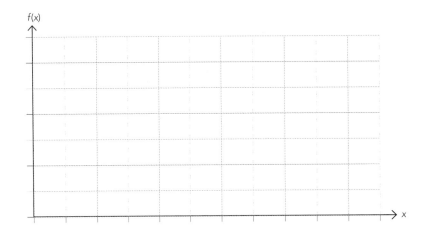

a = _____

b = _____

c = _____

b Calculate the probability that it takes more than 8 minutes to change a battery.

_____

_____

c Calculate the probability that it takes more than 9 minutes to change a battery.

_____

_____

d Calculate the probability that it takes between 6 and 8 minutes to change a battery.

_____

_____

e Calculate the probability that it takes more than 7 minutes to change a battery.

_____

_____

_____

f If two watch batteries are changed, calculate the probability that they both take less than 5 minutes each.

_____

_____

_____

PHOTOCOPYING OF THIS PAGE IS RESTRICTED UNDER LAW. ISBN: 9780170446938

4 The time taken to service a motel room ranges between 12 and 28 minutes, with the most likely time being 16 minutes.

a Draw the probability distribution on the axes below. Label the scales on both axes and show the values of any significant points.

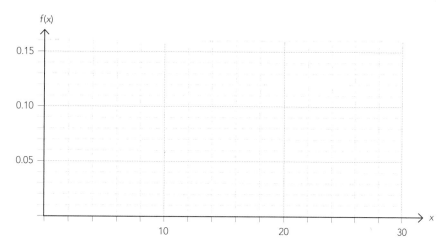

b Write down the parameters for this distribution.

_____

_____

c Calculate the probability that it takes less than 14 minutes to service a room.

_____

_____

d Calculate the probability that it takes between 16 and 20 minutes to service a room.

_____

_____

e Calculate the probability that it takes less than 18 minutes to service a room.

_____

_____

f Above what value do the 20% longest times to service a room lie? (Hints: Call the time taken t; you will end up with a quadratic, which you will need to solve.)

_____

_____

_____

_____

_____

5 The graph below shows the probability distribution for the time in minutes taken by some students to complete a scavenger hunt.

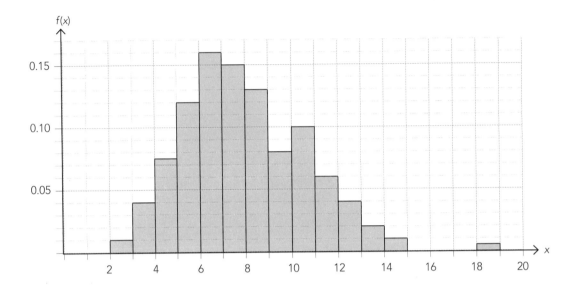

a Write down at least two reasons why the normal distribution would not be appropriate for this distribution.

_____

_____

b The triangular distribution is a possible model for this distribution. Write down the parameters you would use.

_____

_____

c Write down some limitations to the use of a triangular model for this case.

_____

_____

d Using the values on the graph, estimate the probability that $x < 10$.

_____

e Use the triangular distribution and your chosen parameters to estimate the probability that $x < 10$.

_____

_____

f Comment on your answers to **d** and **e**. What does this suggest?

_____

_____

 ISBN: 9780170446938

 Practice questions

Practice question one

a A species of trout (*Salmo statisticus*) lives in a lake. Anglers using the lake were asked to record the number of *Salmo statisticus* caught during each hour they spent fishing.

Number of *Salmo statisticus* caught per hour	Relative frequency
0	0.19
1	0.38
2	0.26
3	0.10
4	0.04
5	0.02
6	0.01

Give the mean and standard deviation for the number of *Salmo statisticus* caught per hour.

_____

_____

_____

b It was suggested that the Poisson distribution would be suitable for this distribution of the number of *Salmo statisticus* caught during each hour spent fishing. Select an appropriate parameter, explain your choice and complete the table.

_____

_____

_____

Number of *Salmo statisticus* caught per hour	Relative frequency	Poisson distribution
0	0.19	
1	0.38	
2	0.26	
3	0.10	
4	0.04	
5	0.02	
6	0.01	

c Discuss whether or not you think the Poisson distribution is a suitable approximation for the distribution of the number of *Salmo statisticus* caught during each hour spent fishing.

_____

_____

_____

_____

_____

_____

_____

d A scientist is attempting to estimate the size of the population of *Salmo statisticus* living in the lake. He did this during the season when nobody is allowed to fish in the lake. This is how he did it:
- In week 1 he captured 625 *Salmo statisticus* using nets, and then he tagged the fish so that they could be recognised.
- These tagged fish were then released back into the lake.
- In week 2, he caught a second group of fish using the same method.
- Out of the 500 fish caught, 39 were tagged.

Use his numbers to estimate the size of the population.

_____

_____

e Explain any statistical assumptions that the scientist made in estimating the population using this method. Give reasons why these assumptions may not be valid.

_____

_____

_____

_____

_____

_____

_____

Practice question two

a The weights of adult female feral cats are normally distributed with a mean of 3.1 kg and a standard deviation of 0.45 kg. Cats that weigh more than 2.5 kg are considered to be a healthy weight. Calculate the percentage of healthy cats that weigh more than 4 kg.

_____

_____

_____

_____

_____

_____

b Much less is known about adult male feral cats. A small survey found that they weighed between 2 and 7 kg, and the most common weight was 3.5 kg.

i Give a reason why the normal distribution would not be a suitable model for use on the distribution of adult male feral cat weights.

_____

_____

_____

ii Suggest a suitable distribution for this data and write down its parameters.

_____

_____

_____

_____

c An adult male and an adult female feral cat are trapped in the same area. Calculate the probability that both weigh less than 3 kg. State and discuss any assumptions that you make.

_____

_____

_____

_____

_____

_____

_____

_____

d Calculate the median weight of male feral cats.

_____

_____

_____

_____

_____

_____

_____

ISBN: 9780170446938

Practice question three

It was estimated that area over which male feral cats range is between 6 and 21 square kilometres, while for female feral cats it is between 2 and 10 square kilometres, depending on available food.

a Using an appropriate probability distribution model, sketch each distribution on the same axes below. Add as much relevant information as possible and clearly label each distribution.

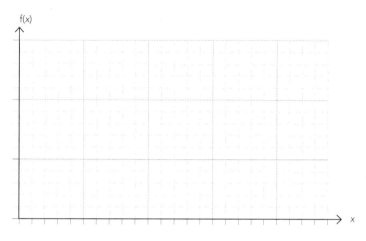

_____

_____

_____

b Justify your use of your chosen probability model.

_____

_____

_____

c A researcher wants to study the 30% of female feral cats with the biggest ranges. Calculate the minimum range for this group of cats.

_____

_____

_____

_____

_____

_____

Practice question four

It is estimated that 7% of owners of domestic cats put bells on the collars of their cats, so that they cannot catch birds so easily.

a If there are 15 cats living in a subdivision, use an appropriate probability distribution model to calculate the probability that more than 2 of the 15 cats wear bells.

b Justify your selection of the probability distribution model used.

c In another subdivision there are 8 cats. The probability that at least one cat wears a bell is 0.513. Calculate the percentage of cats that wear bells in this subdivision. Support your answer with appropriate statistical statements and calculations.

 Answers

Probabilities are rounded to a maximum of 4 dp. Z values are rounded to 3 dp. Answers may vary slightly depending on the method used. Professional judgement should apply.

Discrete vs continuous random variables (pp. 7–8)

Variable	Type of variable
Number of glow worms per m².	Discrete
Jellybean colour	Descriptive
Length of foot	Continuous
Shoe size	Discrete
Time taken to complete a puzzle	Continuous
The cost of an insurance premium	Discrete
Volume of blood in an adult	Continuous
Distance between cities	Continuous
Names of the species of wasps living in New Zealand	Descriptive
Density of wasp nests per hectare	Continuous
Make of car	Descriptive
Mass of a car	Continuous
Car engine capacity	Continuous
Seating capacity of a car	Discrete
Total marks obtained in a quiz	Discrete
Achievement standard result	Descriptive
Frequency of earthquakes	Discrete
Magnitude of an earthquake	Continuous
Height	Continuous
Adult clothing size (e.g. XL)	Descriptive
Current flowing through a wire	Continuous – measured, but could consider electrons as discrete.
Quality of airline meal rated from 1 (poor) to 5 (excellent)	Descriptive. Uses numerals as a code for 'poor' etc.
Age in years	Discrete
Age in seconds	Continuous, although strictly speaking discrete.

The language of probability (pp. 9–11)

Notation	Highlight the numbers included in the description in the left column:	This could also be written as:
$P(X > 5)$	0 1 2 3 4 5 **6 7 8**	$P(X \geq 6)$
$P(X \leq 1)$	**0 1** 2 3 4 5 6 7 8	$P(X < 2)$
$P(0 < X < 4)$	0 **1 2 3** 4 5 6 7 8	$P(1 \leq X \leq 3)$
$P(X \geq 7)$	0 1 2 3 4 5 6 **7 8**	$P(X > 6)$
$P(X < 6)$	**0 1 2 3 4 5** 6 7 8	$P(X \leq 5)$
$P(X = 5)$	0 1 2 3 4 **5** 6 7 8	$P(4 < X < 6)$
$P(1 \leq X \leq 6)$	0 **1 2 3 4 5 6** 7 8	$P(0 < X < 7)$

Words	Highlight the values that X can take:	Notation(s)
$P(X$ is exactly 7$)$	0 1 2 3 4 5 6 **7** 8	$P(X = 7)$ $P(6 < X < 8)$
$P(X$ is greater than 4$)$	0 1 2 3 4 **5 6 7 8**	$P(X > 4)$ $P(X \geq 5)$
$P(X$ is between 0 and 3$)$	0 **1 2** 3 4 5 6 7 8	$P(0 < X < 3)$ $P(1 \leq X \leq 2)$
$P(X$ is less than 2$)$	**0 1** 2 3 4 5 6 7 8	$P(X < 2)$ $P(X \leq 1)$
$P(X$ is at least 3$)$	0 1 2 **3 4 5 6 7 8**	$P(X \geq 3)$ $P(X > 2)$
$P(X$ is between 4 and 7 inclusive$)$	0 1 2 3 **4 5 6 7** 8	$P(4 \leq X \leq 7)$ $P(3 < X < 8)$
$P(X$ is 6 or less$)$	**0 1 2 3 4 5 6** 7 8	$P(X \leq 6)$ $P(X < 7)$
$P(X$ is greater than or equal to 7$)$	0 1 2 3 4 5 6 **7 8**	$P(X \geq 7)$ $P(X > 6)$
$P(X$ is under 3$)$	**0 1 2** 3 4 5 6 7 8	$P(X < 3)$ $P(X \leq 2)$
$P(X$ is over 2$)$	0 1 2 **3 4 5 6 7 8**	$P(X > 2)$ $P(X \geq 3)$
$P(X$ is 4 or more$)$	0 1 2 3 **4 5 6 7 8**	$P(X \geq 4)$ $P(X > 3)$
$P(X$ is more than 1$)$	0 1 **2 3 4 5 6 7 8**	$P(X > 1)$ $P(X \geq 2)$
$P(X$ is not more than 6$)$	**0 1 2 3 4 5 6** 7 8	$P(X \leq 6)$ $P(X < 7)$
$P(X$ exceeds 2$)$	0 1 2 **3 4 5 6 7 8**	$P(X > 2)$ $P(X \geq 3)$
$P(X$ is not less than 6$)$	0 1 2 3 4 5 **6 7 8**	$P(X \geq 6)$ $P(X > 5)$
$P(X$ is less than or equal to 2$)$	**0 1 2** 3 4 5 6 7 8	$P(X \leq 2)$ $P(X < 3)$
$P(X$ is at most 1$)$	**0 1** 2 3 4 5 6 7 8	$P(X \leq 1)$ $P(X < 2)$

Discrete random variables (pp. 13–59)

1 General distributions (pp. 13–30)

Describing distributions (pp. 14–16)

1 Lowest value: 0
 Highest value: 4
 Mode: at 2 with a probability of about 0.41
 Shape: Normal distribution

2 Lowest value: 0
 Highest value: 5
 Mode: modal values at 1 and 5, but only slightly higher than other values
 Shape: Rectangular (or uniform) distribution

3 Lowest value: 0
 Highest value: 6
 Mode: at 4 with a probability of about 0.38
 Shape: Skewed to the left

4 Lowest value: 0
 Highest value: 6
 Modes: at 2 with a probability of 0.35 and at 4 with a probability of 0.25
 Shape: Bimodal distribution

5 Lowest value: 0
 Highest value: 7
 Mode: mode at 3 with a probability of 0.3
 Shape: Irregular distribution

Calculating the mean, standard deviation and variance (pp. 17–21)

1 missing value = 0.3
 $\mu = 1.6$
 $\sigma = 0.9165$
 $\sigma^2 = 0.84$

2 $\mu = 2.5$
 $\sigma = 1.987$
 $\sigma^2 = 3.95$

3 missing value = 0.1
 $\mu = 3.05$
 $\sigma = 1.023$
 $\sigma^2 = 1.047$

4 $\mu = 3.44$
 $\sigma = 1.768$
 $\sigma^2 = 3.126$

5 $\mu = 2$ (Does this surprise you?)
 $\sigma = 1$
 $\sigma^2 = 1$

6 a The winner wins $16, but has paid $2 to enter, so their overall gain is $14.
  The runner-up wins $4, but has paid $2 to enter, so their overall gain is $2. Everybody else pays $2 to enter, so they lose $2.
 b $\mu = (\$14 \times 0.1) + (\$2 \times 0.1) + (-\$2 \times 0.8)$
  $= \$0$

c SD(W) = σ = 4.817
 Var(W) = σ^2 = 23.20

Linear combinations of a discrete random variable: expectation algebra (pp. 22–25)

1 a 2   b 55
  c 100   d 10
  e 16   f 4
  g ?   i 31
  j 36   k 6

2 a E(T) = 1.71   Var(T) = 1.106
  b E(17T) = 29.07 c   Var(17T) = 3.20
  c E(17T + 12) = 41.07 c
   SD(17T + 12) = 17.89 c

3 a E(W) = 2.27   Var(W) = 2.637
  b E(23W + 25) = 77.21 c
   SD(23W + 25) = 37.35 c

4 a E(P) = 2.13   Var(P) = 1.63
  b E(18W + 28) = $66.34
  c SD(18W + 28) = $23.00
  d Changing the fixed nightly rate would make no difference to the standard deviation because it is a fixed amount which does not vary, and it is added to the cost of every cabin.

5 a E(C) = 1.69   Var(C) = 1.294
  b E(23 750C) = $40 137.50
  c SD(23 750C) = $27 016.60

Linear combinations of several discrete distributions: expectation algebra (pp. 26–30)

1 a 22   b 65
  c 55   d 13
  e 3.606   f 10
  g 6   h 7.211
  i 181   j 6.403

2 Yes. Var(X) + Var(Y) = 5.2 + 3.6 = 8.8 = Var(X + Y), so the variables are independent.

3 No. Var(X) + Var(Y) = 1.44 + 0.81 = 2.25.
 Var(X + Y) = 2.1.
 Because Var(X) + Var(Y) ≠ Var(X + Y), the variables are not independent.

4 a If P and Q are independent, then Var(P) + Var(Q) = Var(P + Q).
   ∴ Var(Q) = 1.629.
  b SD(3P + 2Q) = 7.341

5 a E(T) = 1.6
   SD(T) = 1.068
  b Var(S) + Var(T) = 1.793 + 1.141
            = 2.934
   Var(S + T) = 2.95
   Because Var(S) + Var(T) ≠ Var(S + T), S and T are not independent. However, because the difference between 2.934 and 2.95 is

extremely small, for practical purposes S and T are effectively independent.

c $SD(0.5S + 1.2T) \approx 1.45\,(2dp)$

d In doing the calculation, I assumed that S and T were independent. However, because they are only close to being independent, the calculation can only be an estimate.

6 a $E(3P) = \$21.84$
  $SD(3P) = \$5.61$

b $E(4P) = 29.12$
  $SD(4P) = 3.74$

c $E(V) = E(3P + 5M) = \$97.19$
  $Var(V) = Var(3P + 5M) = 3^2 \times 1.87^2 + 5^2 +x$
  $3.97^2 = 425.5$
  $SD(V) = SD(3P + 5M) = \$20.63$

d $E(4P + 4M) = 89.4$
  $Var(4P + 4M) = 4 \times 1.87^2 + 4 \times 3.97^2 =$
  77.03
  $SD(4P + 4M) = 8.777$

e Assumption made: That P and M are independent.
  This may not be a valid assumption because the sales of pumpkins and melons are both likely to be affected by the number of people attending the market: not so many people go to markets on a wet or cold day, so the sales of both would be lower.
  Availability of both pumpkins and melons is likely to depend on weather: warmer weather will probably mean that more of both ripen and become available to sell.
  Check with your teacher if you have a different answer.

7 $E(38T) = 64.98$
  $SD(38T) = 6.483$

8 a $E(C) = \$10.35$
  $SD(C) = \$1.05$

b $E(T) = 2935$
  $SD(T) = 38.66$

c In part **a**, the assumption was made that the random variables P and B were independent.
  • This may not be valid, because both variables will depend on the number wanting breakfast. More wanting breakfast will mean both P and B increase.
  • This may not be valid, because if more choose to have Weet-Bix (B), then fewer will want toast (P).
  Check with your teacher if you have a different answer.

2 The Poisson distribution (pp. 31–45)

Notation	The values that X can take can be highlighted:	Ppd or Pcd?	x value	Answer
$P(X = 4)$	0 1 2 3 **4** 5 6 7 8	Ppd	4	0.0471
$P(X \leq 1)$	**0 1** 2 3 4 5 6 7 8	Pcd	1	0.5578
$P(X > 2)$	0 1 2 **3 4 5 6 7 8**	Pcd	2	$1 - P(X \leq 2)$ $= 0.1912$
$P(X < 3)$	**0 1 2** 3 4 5 6 7 8	Pcd	2	0.8088
$P(X \geq 4)$	0 1 2 3 **4 5 6 7 8**	Pcd	3	$1 - P(X \leq 3)$ $= 0.0656$

1

	Values **x** can take	Probability
$P(x = 0)$	**0** 1 2 3 4 5 6 7 8	**0.4066**
$P(x = 3)$	0 1 2 **3** 4 5 6 7 8	0.0494
$P(x = 5)$	0 1 2 3 4 **5** 6 7 8	0.0020
$P(x < 2)$	**0 1** 2 3 4 5 6 7 8	0.7725
$P(x > 1)$	0 1 **2 3 4 5 6 7 8**	0.2275
$P(x \geq 3)$	0 1 2 **3 4 5 6 7 8**	0.0628
$P(1 < x < 4)$	0 1 **2 3** 4 5 6 7 8	0.2141
$P(x \leq 4)$	**0 1 2 3 4** 5 6 7 8	0.9977
$P(4 < x \leq 7)$	0 1 2 3 4 **5 6 7** 8	0.0023

The last two answers add to 1 because x has to take a value between 0 and 8 inclusive. In theory x could be greater than 8, but the probabilities of all the numbers above 8 add to less than 0.0001.

ISBN: 9780170446938

2

	Values x can take	Probability
$P(x = 0)$	0	0.1108
$P(x = 3)$	3	0.1966
$P(x = 10)$	10	0.0001
$P(x < 3)$	0, 1, 2	0.6227
$P(x \geq 7)$	7, 8, 9, 10, …	0.0075
$P(4 < x < 7)$	5, 6	0.065
$P(5 < x \leq 8)$	6, 7, 8	0.0244
$P(x > 1)$	2, 3, 4, 5, 6, 7, 8, 9, 10, …	0.6454
$P(x \leq 1)$	0, 1	0.3546

$\mu = 2.2$
$\sigma = 1.483$

Applications of the Poisson distribution (pp. 37–43)

1
  a 0.5488   **b** 0.1185
  c 0.0034   **d** 0.8781
  e $\sigma = 0.7746$ $\sigma^2 = 0.6$
  f 6 cyclones   **g** 0.1606

  h Any four of:
- Cyclones are discrete events, which occur in a continuous interval of time.
- Cyclones are relatively rare events.
- Cyclones occur randomly and unpredictably. This may not be quite true, though, because for many islands there is a cyclone season when they would be more likely.
- Cyclones occur independently of each other.
- Cyclones cannot occur simultaneously.

2
  a 0.3614
  b 0.1127
  c 0.9984
  d 0.1205
  e 0.1205
  f Six earthquakes
  g 0.1190
  h Earthquakes do not always occur randomly. In the period following a big earthquake, there are usually aftershocks, which may be over four on the Richter scale.

3
  a 0.0613
  b 0.0803
  c 0.9197
  d 'At least three jars' $\Rightarrow$ 3, 4, 5, 6, 7 or more jars.
'No more than two jars' $\Rightarrow$ 0, 1 or 2 jars. Altogether that means 0 jars or more, which is certain so the probability is 1. 0.0803 + 0.9197 = 1, so these are complementary events.
  e Four jars
  f 0.0053
  g 0.2381
  h **1** Each occurrence is independent of others. Each sale of a jar of snails is probably independent of others, but if a customer sees another buying snails, they may be more likely to buy them too.
    2 Events must not occur simultaneously. This condition probably doesn't apply because customers may buy several jars at once.
    3 Events must occur randomly and unpredictably. Buying of jars of snails is likely to be random and unpredictable, unless they are commonly bought at a particular time of the year for festive or other reasons.
    4 For a small interval, the probability of an event occurring is proportional to the size of the interval. This should apply to selling jars of snails, as long as the demand is consistent throughout the year.

4
  a

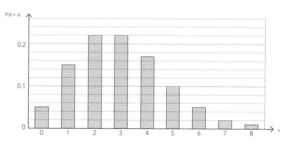

  b Any two of:
Discrete distribution – X can take only whole numbers – they can't sell half a pot of jellied eel.
Unimodal – modes at $X = 2$ and $X = 3$, which are adjacent. The most likely

number of pots sold is two or three in a week.
Skewed to the right – Poisson is the distribution of rare events, so the delicatessen is most likely to sell three pots or fewer each week.

c 0.4232
d 0.8008
e 6 pots
f 0.1528
g 0.0302

5 a $\mu = 2.57$ (allow some variation)
b Use $\lambda = 2.6$

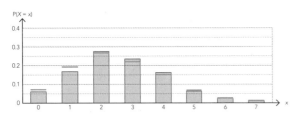

c 0.4686
d 0.0055
e 0.4062
f Graph shows probabilities from 0 to 7 only, while the Poisson distribution for $\lambda = 2.6$ has values between 0 and 11. However, this may be because the probabilities for 8 and above (0.0038, etc.) are so small they would not show on the graph.
Both graphs are unimodal with a mode at $x = 2$. Both are skewed to the right. The probabilities for the Poisson distribution are slightly higher when $x = 1$, and slightly lower for $x = 2$, but otherwise the graphs are very similar, so the Poisson distribution is a good model for the distribution of wetas.

6 a 0.67
b 0.34
c $P(X = 0) = 0.51$, so most areas of one hectare have no nests, which makes the nests relatively rare. The Poisson distribution is used for rare events.
d $\mu = 1.33$ nests per hectare, $\sigma = 1.153$

e

n	0	1	2	3	4	5	6
P(N = n)	0.51	0.16	0.10	0.08	0.06	0.05	0.04
	0.2645	0.3518	0.2339	0.1037	0.0345	0.0092	0.0020

f The mode for the actual distribution is much too high (0.51) compared with the mode for the theoretical distribution (0.3518). It is also in the wrong place – at 0 nests in the actual distribution and at one nest for the theoretical distribution.
The actual distribution probabilities are much too high for no nests, too low for one and two nests, about right for three nests and too high for four, five and six nests per hectare.
For the Poisson distribution:
mean = variance
$\mu = \lambda = 1.33$
$\sigma^2 = (1.778)^2 = 3.161$
$\therefore$ mean $\neq$ variance
For all the reasons above, the Poisson distribution is not a very good model for this distribution of wasp nests.

Inverse Poisson problems (pp. 44–45)
1 0.8 cyclones per year
2 1.31 earthquakes per year
3 0.58 jars per week
4 0.4308 pots per week $\Rightarrow$ 0.8616 pots in two weeks
5 1.02 weta per square metre
$\Rightarrow$ 4.08 weta per 4 square metres
P = 0.4180
6 0.439 wasps nests per hectare
$\Rightarrow$ 2.195 wasps nests per 5-hectare block
P = 0.6442

3 Binomial distribution (pp. 46–59)
1

	Values x can take	Probability
P(x = 0)	**0** 1 2 3 4 5 6	**0.5314**
P(x = 2)	0 1 **2** 3 4 5 6	0.0984
P(x = 6)	0 1 2 3 4 5 **6**	0
P(x < 2)	**0 1** 2 3 4 5 6	0.8857
P(x > 1)	0 1 **2 3 4 5 6**	0.1143
P(x ≥ 3)	0 1 2 **3 4 5 6**	0.0159
P(1 < x < 4)	0 1 **2 3** 4 5 6	0.113
P(x ≤ 3)	**0 1 2 3** 4 5 6	0.9987
P(3 < x ≤ 6)	0 1 2 3 **4 5 6**	0.0013

The last two answers add to 1 because x has to take a value between 0 and 6 inclusive.

2

	Values x can take	Probability
P(x = 0)	0	0.0060
P(x = 8)	8	0.0106
P(x = 10)	10	0.0001
P(x < 3)	0, 1, 2	0.1673
P(x ≥ 4)	4, 5, 6, 7, 8, 9, 10	0.6178
P(3 < x < 8)	4, 5, 6, 7	0.6055
P(5 < x ≤ 10)	6, 7, 8, 9, 10	0.1662
P(x > 1)	2, 3, 4, 5, 6, 7, 8, 9, 10	0.9537
P(x ≤ 1)	0, 1	0.0463

$\mu = 4$
$\sigma = 1.549$

3 a 0.1719
  b 0.4769
  c 0.0885
  d 0.2568
  e 0.2201
  f There are only two possible outcomes: buying tropical or orange Juicies.
  There is a fixed number of trials: consider eight sales of Juicies.
  The probability stays constant: cannot be certain of this — flavours may become more or less popular over time.
  The outcome of each trial is independent of other outcomes: what one person buys probably doesn't affect what the next person buys, but sometimes people copy others.

4 a 0.1615
  b 0.7752
  c 0.2247
  d The events 'getting zero, one or two sixes' and 'getting three or more' are complementary — one of those must happen. Consequently their probabilities should add to 1. (However, 0.7752 + 0.2247 add to 0.9999. This means that the probabilities of getting eight, nine or ten sixes add to 0.0001.)
  e $\mu = 1.\dot{6}$, $\sigma = 1.1785$
  f 0.1043
  g There are only two possible outcomes: throwing a six or not throwing a six.
  There is a fixed number of trials: 10 throws of the die.

The probability stays constant: the same die is used for each throw.
The outcome of each trial is independent of other outcomes: the outcome of each throw of the die is not influenced by what has been thrown in earlier throws.

5 a 0.0687
  b 0.7945 or 0.7946
  c $\mu = 2.4$, $\sigma = 1.386$
  d 0.0425
  e 0.5618
  f There are only two possible outcomes: getting 'orange' or 'not orange' chews (or 'raspberry or passionfruit' or 'not raspberry or passionfruit'-flavoured chews).
  There is a fixed number of trials: packets contain exactly 12 chews.
  The probability stays constant: have to take the manufacturer's word that there are equal probabilities of getting each flavour.
  The outcome of each trial is independent of other outcomes: need to assume that the different flavours occur randomly in each packet. It is possible that the chews of the same flavour stick together, so more of one flavour than expected could occur in a packet.

6 a 0.2437
  b 0.3529
  c 0.0057
  d

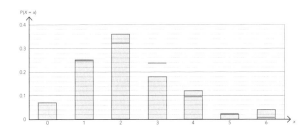

  e Both graphs have a mode at 2, and are skewed to the right.
  The model fits well for 0, 1 or 5 bunks occupied. However, the actual occupancy for 2, 4 and 6 bunks is higher than predicted by the binomial distribution. This may be because travellers who would be likely to use the cabins, often travel as couples. The actual occupancy for 3 is lower than the binomial model predicts. Therefore, the binomial distribution is not a particularly good one for predicting bunk occupancy in his cabins.

7 a

Number correct	0	1	2	3	4	5	6
Number of students	2	9	6	6	4	1	0
Probability	0.0714	0.3214	0.2143	0.2143	0.1429	0.0357	0
Binomial distribution	0.1780	0.3560	0.2966	0.1318	0.0330	0.0044	0.0002

b There are only two possible outcomes: computer or chimpanzee gets answer right or wrong.
There is a fixed number of trials: exactly six questions.
The probability stays constant: random selection of answers should mean that the probability of getting each question correct is 0.25.
The outcome of each trial is independent of other outcomes: must assume that whichever answer is selected for a question does not affect which answer is selected for the next question.

c In table above.

d Yes. The probabilities that students got 0, 1 and 2 out of 6 were lower than the binomial distribution would predict.
So more students than the binomial distribution would predict got 3, 4 or 5 out of 6.
Students' mean = 0.3214 + (2 × 0.2143) + (3 × 0.2143) + (4 × 0.1429) + (5 × 0.0357) = 2.143
Binomial mean = np = 6 × 0.25 = 1.5
∴ the students' mean is considerably higher than mean for the binomial distribution
∴ the students definitely did better than the chimpanzees.

8 The expected number of sixes in 24 throws
$$= 24 \times \frac{1}{6} = 4.$$
Getting just two more sixes than expected in 24 throws is not very surprising.
For n = 24 and $p = \frac{1}{6}$, P(X = 6) = 0.1084.
This is not a very low probability, so he could get 6 sixes simply by chance.
24 trials is not very many. If he wanted to demonstrate more convincingly that his dice was weighted, he should do lots more trials.

9 a

	0	1	2	3	4	5
Binomial	0.5987	0.3151	0.0743	0.0105	0.0010	0.0001
Poisson	0.6065	0.3033	0.0758	0.0126	0.0016	0.0002

b $\mu = np = 0.5$

c $\lambda = 0.5$

d The distributions are very similar because a binomial probability of 0.05 means that a fault is a rare event, which is a characteristic of the Poisson distribution.

Inverse binomial problems (pp. 57–59)

1 a $(1 - p)^6 = 0.7730 \Rightarrow$ 4.2% broken
 b $(1 - p)^6 = 0.87 \Rightarrow$ 2.3% broken
 c $(0.958)^n = 0.87 \Rightarrow n = 3.246$
 ∴ three in each pack ⇒ 12.1% returned

2 $(1 - p)^{20} = 0.39 \Rightarrow$ P(plant thrives) = 0.9540

3 $(0.5)^n = 0.05 \Rightarrow n = 4.321$
 ∴ she needs to toss the coin five times ⇒ P(more than one head) = 0.9686

4 $(0.9)^n = 0.\dot{3} \Rightarrow n = 10.43$
 ∴ he needs to put 11 into each pack ⇒ 68.62% will have at least one red jellybean.

Continuous random variables (pp. 60–108)

1 The normal distribution (pp. 60–94)
Sketching normal distribution (pp. 61–62)

1
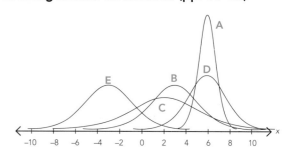

2
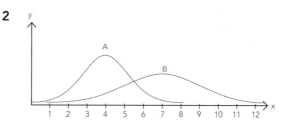

3
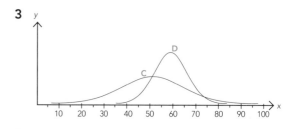

4
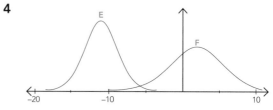

ISBN: 9780170446938

The standard normal distribution (pp. 63–65)

1 a 2   b −3
2 a 0   b −3.5
3 a 1.381   b −3.238
4 a −2.581   b 0.6452
5 a 1.375   b −2.5

Reading normal distribution tables (pp. 66–67)

1 0.4821   2 0.4066
3 0.2707   4 0.3497
5 0.4750   6 0.5000

Calculating probabilities (pp. 66–74)

Calculating a probability on the right side of the curve
$Z = 2 \Rightarrow p = 0.4772$

Probabilities below a value
$Z = 2 \Rightarrow p = 0.9772$

Probabilities from the right tail of the curve
$Z = 1.\dot{3} \Rightarrow p = 0.0913$

Probabilities on the left of the curve
$Z = 2.\dot{4} \Rightarrow p = 0.4927$

Probabilities from a tail on the left side of the curve
$Z = 2.\dot{4}\, p = 0.0072$

Probabilities from both sides of the curve
$Z = 1.\dot{3}$ and $0.\dot{8} \Rightarrow p = 0.7218$

Probabilities from two tails of the curve
$Z = 2$ and $0.\dot{2} \Rightarrow p = 0.4349$

Calculating probabilities by difference
$Z = 1.\dot{7}$ and $0.\dot{8} \Rightarrow p = 0.1492$

Mixing it up (pp. 76–80)

1 a 58.34%
  b 0.6969
  c 0.0290 or 0.0291
  d 0.9144 x 50 000 = 45 720
  e The normal distribution is probably suitable in this situation because:
    The data is continuous: weight is a measured variable.
    Most data is clustered around a central value, with a few extreme values either side (bell-shaped curve): we cannot be certain of this without seeing a graph of the distribution, but that is usually the case with weights of animals.
    There should be no upper or lower limit to the values it can take: 0 is not a possible weight for a tuatara, and there is no upper limit to what one can weigh.
    The data is symmetrical: we cannot be certain of this — it may be skewed to the left if there are lots of young tuataras.

2 a 0.0043
  b 5.71 %
  c $p = 0.644 \Rightarrow 772$ balls or $p = 0.643 \Rightarrow 772$ balls
  d $0.1464^2 = 0.0214$
  e The normal distribution is probably suitable in this situation because:
    The data is continuous: length of string is a measured variable.
    Most data is clustered around a central value, with a few extreme values either side (bell-shaped curve): we cannot be certain of this without seeing a graph of the distribution, but it is likely for the variation in the lengths of string.
    There should be no upper or lower limit to the values it can take: the lengths of string can take any value, so this applies.
    The data is symmetrical: we cannot be certain of this without seeing the probability distribution, but it is likely.

3 a 63.65% or 63.64%
  b 0.9107
  c $0.9818^3 = 0.9464$
  d $p = 0.2577 \Rightarrow 33$ samples
  e The normal distribution may not be suitable in this situation because:
    The data is not continuous because number of pollen grains is discrete data. In theory this means that the normal distribution is not appropriate, but in practice it is useful in this situation because the range in the number of pollen grains is so large.
    Most data is clustered around a central value, with a few extreme values either side (bell-shaped curve): we cannot be certain of this without seeing a graph of the distribution.The data is symmetrical: we cannot be certain of this without seeing a graph of the distribution.

4 a 55.4
  b $P(X \geq 70) = 0.18$
  c $P(X \geq 70) = 0.1881$
  d Graph: $P(X \leq 40) = 0.18$
    Normal distribution (by symmetry): $P(X \leq 40) = 0.1749$
  e The normal distribution is not suitable for use with this data because:
    1 The data is not continuous because number of marks is discrete data.

2 For the normal distribution to apply, there should be no upper or lower limit to the values it can take. However, the percentages have a lower limit of 0 and an upper limit of 100.
The data is suitable for use with the normal distribution because:
It is symmetrically clustered around a central value, with a few extreme values either side (bell-shaped curve).
The calculations for P(X ≥ 70) and P(X ≤ 40) using the graph and the normal distribution are quite close.

5 For: Continuous data (length)
No upper or lower limit
Most data is clustered around a central value, with a few extreme values either side
Bell-shaped curve, but skewed
Against: Not symmetrical

6 For: Continuous data (length)
No upper or lower limit
Most data is clustered around a central value, with a few extreme values either side
Symmetrical
Bell-shaped curve
Against: None

7 For: Continuous data (time)
No upper or lower limit
Most data is clustered around a central value
Symmetrical
Against: Not a bell-shaped curve — triangular distribution

8 For: Most data clustered around a central value, with a few extreme values either side
Bell-shaped curve
Symmetrical
Against: Data not continuous (number of credits is discrete)
Lower limit of 0 and upper limit depending on the maximum number of credits offered.

Inverse normal distribution (pp. 81–86)
Using a probability to find Z
1 2.62 **2** 0.12
3 3.06 **4** 1.301
5 1.653 **6** 0.772
7 2.195 **8** 0.368
9 1.983 or 1.984 **10** 2.546 or 2.547

Using Z to calculate the value of X – where Z is positive
$Z = 0.524 \Rightarrow$ weight = 4.336 kg

Using Z to calculate the value of X – where Z is negative
$Z = -1.282 \rightarrow$ weight = 3.523 kg

Using Z to calculate the mean
$Z = 1.645 \Rightarrow \mu = 4.33$ kg

Using Z to calculate the standard deviation
$Z = 0.674 \Rightarrow \sigma = 0.4451$ kg

Mixing it up (pp. 87–88)
1 **a** $\mu = 2.2$ kg **b** $\sigma = 0.22$
c LQ = 2.245, UQ = 2.555
d $\sigma = 0.280$ **e** $\mu = 3.3$ kg
2 **a** $\sigma = 0.75$ **b** $x = 3.53$ km
c $\mu = 11.2$ km **d** $x = 10.0$ km
e $\sigma = 1.45$ km

Mixed normal distribution problems (pp. 89–90)
1 **a** 0.1151
b 28.81%
c 15.42 ⇒ 15 or 16
d 2.737 kg
e 0.3005
f 2.89 kg
2 **a** 0.4452
b 2.28%
c 146
d 1.565 kg
e 1.371 kg to 1.630 kg
f 0.1402

Continuity correction (pp. 91–94)
See tables opposite on page 125)

2 The rectangular (continuous uniform) distribution (pp. 95–101)

1 **a** $f(x) = \dfrac{1}{12 - 4} = 0.125$

b

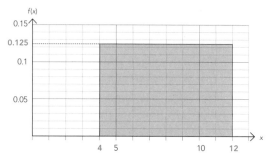

c $p = \dfrac{9 - 6}{8} = 0.375$

Continuity correction (pp. 84–87)

Words	Highlight the values that X can take:	Sketch of area	X value(s)	Probability
P(X is greater than 5)	0 1 2 3 4 5 **6 7 8**		5.5	0.0668
P(X is at least 5)	0 1 2 3 4 **5 6 7 8**		4.5	0.3085
P(X is less than 4)	**0 1 2 3** 4 5 6 7 8		3.5	0.3085
P(X is between 4 and 7)	0 1 2 3 4 **5 6** 7 8		4.5 and 6.5	0.3023
P(X is 5)	0 1 2 3 4 **5** 6 7 8		4.5 and 5.5	0.2417
P(X is 6 or less)	**0 1 2 3 4 5 6** 7 8		6.5	0.9938
P(X is over 2)	0 1 2 **3 4 5 6 7 8**		2.5	0.9332
P(X is between 2 and 5 inclusive)	0 1 **2 3 4 5** 6 7 8		1.5 and 5.5	0.9270

Situation	Continuity correction ✓ or ✗	Sketch of area	X value(s)	Probability
A test is marked out of 50. The marks are normally distributed with a mean of 31.4 and a standard deviation of 5.2. Calculate the percentage of marks that lie between 25 and 40.	✓ (Discrete data)		24.5 and 40.5	86.77%
A group of teenagers have their ages recorded to the nearest year. The ages are normally distributed with a mean of 16.2 years and a standard deviation of 1.1 years. Calculate the probability that the recorded age of a teenager is more than 16.	✓ (Rounded data)		16.5	0.3923
A group of adults have their ages recorded to the nearest year. The ages are normally distributed with a mean of 56.8 years and a standard deviation of 12.1 years. Calculate the probability that the recorded age of an adult is less than 45.	✗ (Rounded but range too large)		45 (with cc: 44.5)	0.1647 (with cc: 0.1547)
The times taken for the group of adults to complete a task on a website are normally distributed with a mean of 26.8 minutes and a standard deviation of 5.4 minutes. Calculate the probability that an adult completed the task in less than 30 minutes.	✗ (Continuous data)		30	0.7233
Beck has a 48% success rate when she attempts to shoot netball goals from the edge of the circle. If she has 32 attempts, use the normal distribution to estimate the probability that she is successful in at least 14 of her attempts. (Hint: For binomial, $\mu = np$ and $\sigma = \sqrt{np(1-p)}$)	✓ (Number of successes is discrete)		13.5	0.7448
The number of items in the pencil cases of a group of students is normally distributed with a mean of 9.4 and a standard deviation of 2.6. Calculate the probability that a student has fewer than six or more than 12 items in their pencil case.	✓ (Discrete data)		5.5 and 12.5	0.0667 + 0.1166 = 0.1833

2 a $a = 10, b = 25$

b $f(x) = 0.0\dot{6}$

c $p = \dfrac{14 - 11}{15} = 0.2$

3 a

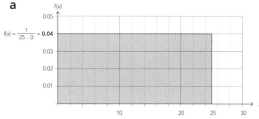

$f(x) = \dfrac{1}{25 - 0} = \textbf{0.04}$

b $\dfrac{4}{25} = 0.16$

c $\dfrac{3}{25} = 0.12$

d

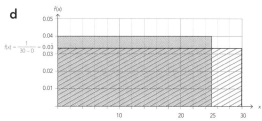

$f(x) = \dfrac{1}{30 - 0} = 0.0\dot{3}$

e $\dfrac{4}{30} = 0.1\dot{3}$

f $\dfrac{30}{2} = 15$ m

4 a

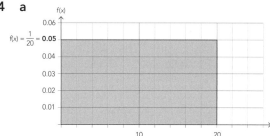

$f(x) = \dfrac{1}{20} = \textbf{0.05}$

b 0.3

c 0.125

d $(0.25)^2 = 0.0625$

e Every 18 minutes

f $0.0\dot{5}$

5 a

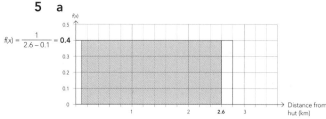

$f(x) = \dfrac{1}{2.6 - 0.1} = \textbf{0.4}$

b 0.04

c 1.35 km

d

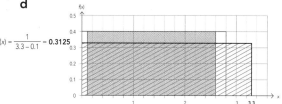

$f(x) = \dfrac{1}{3.3 - 0.1} = \textbf{0.3125}$

e $\dfrac{0.4 + 0.5}{3.2} = 0.28125$

f $\dfrac{0.3}{2.5} \times \dfrac{0.5}{3.2} = 0.01875$

6 a Use 24-hour clock time

b $a = 8, b = 18$

c 0.46

d 0.5

e The rectangular distribution should be flat across the top, but this distribution of phone calls dips a little in the middle, and is higher at the ends. This is supported by comparing the actual proportion of phone calls between 11 a.m. and 4 p.m. (0.46) compared with the theoretical number, based on a rectangular model (0.5).

3 The triangluar distribution (pp. 102–108)

1 a $a = 1, \ b = 6, \ c = 3$

b $f(x) = \dfrac{2}{b - a}$

$= \dfrac{2}{6 - 1}$

$= 0.4$

c height $= f(5) = \dfrac{2(b - x)}{(b - a)(b - c)}$

$= \dfrac{2(6 - 5)}{(6 - 1)(6 - 3)}$

$= \dfrac{2}{5 \times 3}$

$= 0.1\dot{3}$

d

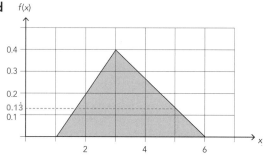

e Probability = $\frac{1}{2}$ x base x height

$$= \frac{1}{2} \times 1 \times 0.1\dot{3}$$

$$= 0.0\dot{6}$$

f $f(2.5) = \frac{2(x-a)}{(b-a)(c-a)}$

$$= \frac{2(2.5-1)}{(6-1)(3-1)}$$

$$= \frac{3}{5 \times 2}$$

$$= 0.3$$

g Area of a trapezium $= \frac{\text{sum of parallel sides}}{2}$

x vertical height

$$= \frac{0.3 + 0.4}{2} \times 0.5$$

$$= 0.175$$

2 a $f(16) = 0.1$

b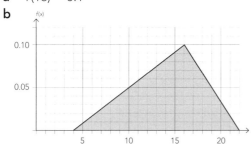

c $f(10) = 0.05$
d $P(x < 10) = 0.15$
e $f(18) = 0.075$
f $P(16 < x < 18) = 0.175$
g Height = $f(12) = 0.0\dot{6}$
  $P(x < 12) = 0.2\dot{6}$
  Total area $= 1 - 0.2\dot{6} = 0.7\dot{3}$

3 a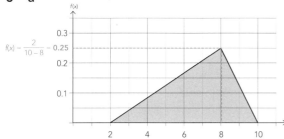

b $h = 0.25$; $p = 0.25$
c $h = 0.125$; $p = 0.0625$

d $h = 0.1\dot{6}$; $p = 0.41\dot{6}$
e $h = 0.208\dot{3}$; $p = 0.4792$
f $h = 0.125$; $p = (0.1875)^2 = 0.03516$

4 a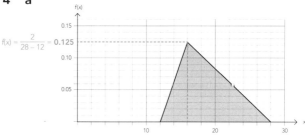

b $a = 12$, $b = 28$, $c = 16$
c $h = 0.0625$; $p = 0.0625$
d $h = 0.08\dot{3}$; $p = 0.41\dot{6}$
e $h = 0.1041\dot{6}$; $p = 0.4792$
f $t = 21.8$ minutes or 34.2 minutes. 34.8 minutes lies outside the range, so $t = 21.8$ minutes.

5 a 1 The distribution has a lower limit of zero, which a normal distribution should not have.
  2 The distribution is not symmetrical about the mode: it is skewed to the right.
  3 It is not really bell-shaped.
b $a = 2$, $b = 15$, $c = 7$
c 1 The low probability between 9 and 10, and the peak between 10 and 11 don't fit well.
  2 The outlier(s) between 18 and 19 need to be ignored in order to make it fit.
d 0.7650
e $h = 0.0.0962$, 0.7596
f The two values are close, indicating that the triangular model is a good fit for this data.

Practice questions (pp. 109–114)

Practice question one (pp. 109–110)

a $\mu = 1.52$
  $\sigma = 1.229$
b Let $\lambda = 1.52$
  I chose 1.52 because λ is the mean of a Poisson distribution.

Number of *Salmo statisticus* caught per hour	Relative frequency	Poisson distribution
0	0.19	0.2187
1	0.38	0.3324
2	0.26	0.2527
3	0.10	0.1280
4	0.04	0.0486
5	0.02	0.0148
6	0.01	0.0037

c Suitable because:

1 Fit is reasonably good – predicted frequencies for Poisson model a little low for anglers who caught one fish per hour, but the rest are reasonably close.

2 If the probability of catching 7 fish in an hour (0.0008) is included, the Poisson frequencies add to 0.997, which is very close to 1.

3 The model has about the same range than the actual data. The probability of catching 7 fish in an hour is very low (0.0008).

4 For a Poisson distribution $\lambda = \sigma^2$. In this case $\lambda = 1.52$ and $\sigma^2 = 1.51$. These are very close, which suggests that the Poisson distribution is a good fit for the data.

d $\dfrac{\text{Number of tagged fish in week 1}}{\text{Total population}} = \dfrac{\text{Number of tagged fish in week 2}}{\text{Total number caught in week 2}}$

$\therefore \dfrac{625}{\text{Total population}} = \dfrac{39}{500}$

Estimate of total trout population = 8013

e 1 He assumed that after he had returned his tagged fish to the lake in week 1, the proportion of tagged fish in the population remained the same during week 2. This might not be valid because the tagged fish might be more likely to die, or be eaten by a predator.

2 In week 2, he assumed that the probability of catching a tagged fish was the same as the probability of catching an untagged fish.
This might not be valid because the tagged fish might be more or less likely to be caught in week 2.

Practice question two (pp. 111–112)

a $P(x > 2.5) = 0.90878$
$P(x > 4) = 0.02275$
$\therefore 2.5\%$

b i Distribution is not symmetrical — to be symmetrical the most common weight would need to be 4.5 kg.

ii Triangular distribution with $a = 2$, $b = 7$ and $c = 3.5$.

c $P(\text{male} < 3\text{ kg}) = 0.1\dot{3}$
$P(\text{female} < 3\text{ kg}) = 0.41207$
$P(\text{both} < 3\text{ kg}) = 0.05494$
Assumption: The events are independent. This may not be the case, because they live in the same area, so may be closely related and therefore either big or small. In addition, their size may be determined by how much food there is, which could also cause both to be big or small.

d Let m = median, h = height of graph at median.

$$h = \frac{2.8 - 0.4\,m}{3.5}$$

$$P(x > m) = \frac{1}{2} \times (7 - m) \times h = 0.5$$

Substitute h and solve $\rightarrow m = 4.0419$ (or 9.95).
$\therefore$ median weight is 4.0420 kg.

Practice question three (pp. 113)

a

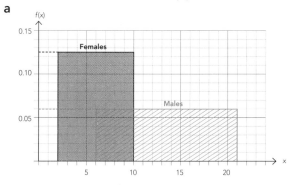

b A maximum and minimum value are given, but no modal value. As far as we know, all probabilities are equally likely.

c $(10 - x) \times 0.125 = 0.3$, $\rightarrow x = 7.6$ square kilometres

Practice question four (pp. 114)

a Binomial, $n = 15$, $p = 0.07 \rightarrow p = 0.08287$
(not Poisson because you do not have
discrete events in a continuous interval)

b Binomial:

Just two outcomes – either a cat wears a bell
or it doesn't.

Fixed number of trials – 15 cats.

Fixed probability – 7%.

Trials independent – may not be the case
because several cats could belong to one
owner in which case those cats would all be
more or less likely to wear bells.

c $P(c = 0) = 0.487$, $1 - p = 0.914 \therefore 8.6\%$ wear
bells.

Pull-out section (pp. 133–135)

A Poisson; $\lambda = 1.2$, $p = 0.3012$

B Triangular; $a = 2$, $b = 20$, $c = 15$, $p = 0.7265$

C Normal; $\mu = 17.4$, $\sigma = 0.6$; $p = 0.5889$

D Binomial; $n = 12$, $\pi = 0.55$, $p = 0.1345$

E Triangular; $a = 0$, $b = 12$, $c = 2$. $p = 0.125$

F Normal; $\mu = 53$, $\sigma = 3$; $p = 0.2525$

G Binomial; $n = 7$, $\pi = 0.18$, $p = 0.1154$

H Poisson; $\lambda = 3.1$. $p = 0.2018$

I Rectangular; $a = 0$, $b = 6$, $p = 0.25$

J Rectangular; $a = 20$, $b = 270$, $p = 0.32$

Binomial Distribution

Each entry gives the probability that a binomial random variable X, with the parameters n and π, has the value x.

$$P(X = x) = \binom{n}{x}\pi^x(1 - \pi)^{n-x}$$

$$\mu = n\pi, \qquad \sigma = \sqrt{n\pi(1 - \pi)}$$

n	x	0.05	0.1	0.15	1/6	0.2	0.25	0.3	1/3	0.35	0.4	0.45	0.5
4	0	0.8145	0.6561	0.5220	0.4823	0.4096	0.3164	0.2401	0.1975	0.1785	0.1296	0.0915	0.0625
	1	0.1715	0.2916	0.3685	0.3858	0.4096	0.4219	0.4116	0.3951	0.3845	0.3456	0.2995	0.2500
	2	0.0135	0.0486	0.0975	0.1157	0.1536	0.2109	0.2646	0.2963	0.3105	0.3456	0.3675	0.3750
	3	0.0005	0.0036	0.0115	0.0154	0.0256	0.0469	0.0756	0.0988	0.1115	0.1536	0.2005	0.2500
	4		0.0001	0.0005	0.0008	0.0016	0.0039	0.0081	0.0123	0.0150	0.0256	0.0410	0.0625
5	0	0.7738	0.5905	0.4437	0.4019	0.3277	0.2373	0.1681	0.1317	0.1160	0.0778	0.0503	0.0313
	1	0.2036	0.3281	0.3915	0.4019	0.4096	0.3955	0.3602	0.3292	0.3124	0.2592	0.2059	0.1563
	2	0.0214	0.0729	0.1382	0.1608	0.2048	0.2637	0.3087	0.3292	0.3364	0.3456	0.3369	0.3125
	3	0.0011	0.0081	0.0244	0.0322	0.0512	0.0879	0.1323	0.1646	0.1811	0.2304	0.2757	0.3125
	4		0.0005	0.0022	0.0032	0.0064	0.0146	0.0284	0.0412	0.0488	0.0768	0.1128	0.1563
	5			0.0001	0.0001	0.0003	0.0010	0.0024	0.0041	0.0053	0.0102	0.0185	0.0313
6	0	0.7351	0.5314	0.3771	0.3349	0.2621	0.1780	0.1176	0.0878	0.0754	0.0467	0.0277	0.0156
	1	0.2321	0.3543	0.3993	0.4019	0.3932	0.3560	0.3025	0.2634	0.2437	0.1866	0.1359	0.0938
	2	0.0305	0.0984	0.1762	0.2009	0.2458	0.2966	0.3241	0.3292	0.3280	0.3110	0.2780	0.2344
	3	0.0021	0.0146	0.0415	0.0536	0.0819	0.1318	0.1852	0.2195	0.2355	0.2765	0.3032	0.3125
	4	0.0001	0.0012	0.0055	0.0080	0.0154	0.0330	0.0595	0.0823	0.0951	0.1382	0.1861	0.2344
	5		0.0001	0.0004	0.0006	0.0015	0.0044	0.0102	0.0165	0.0205	0.0369	0.0609	0.0938
	6					0.0001	0.0002	0.0007	0.0014	0.0018	0.0041	0.0083	0.0156
7	0	0.6983	0.4783	0.3206	0.2791	0.2097	0.1335	0.0824	0.0585	0.0490	0.0280	0.0152	0.0078
	1	0.2573	0.3720	0.3960	0.3907	0.3670	0.3115	0.2471	0.2048	0.1848	0.1306	0.0872	0.0547
	2	0.0406	0.1240	0.2097	0.2344	0.2753	0.3115	0.3177	0.3073	0.2985	0.2613	0.2140	0.1641
	3	0.0036	0.0230	0.0617	0.0781	0.1147	0.1730	0.2269	0.2561	0.2679	0.2903	0.2918	0.2734
	4	0.0002	0.0026	0.109	0.0156	0.0287	0.0577	0.0972	0.1280	0.1442	0.1935	0.2388	0.2734
	5		0.0002	0.0012	0.0019	0.0043	0.0115	0.0250	0.0384	0.0466	0.0774	0.1172	0.1641
	6			0.0001	0.0001	0.0004	0.0013	0.0036	0.0064	0.0084	0.0172	0.0320	0.0547
	7						0.0001	0.0002	0.0005	0.0006	0.0016	0.0037	0.0078
8	0	0.6634	0.4305	0.2725	0.2326	0.1678	0.1001	0.0576	0.0390	0.0319	0.0168	0.0084	0.0039
	1	0.2793	0.3826	0.3847	0.3721	0.3355	0.2670	0.1977	0.1561	0.1373	0.0896	0.0548	0.0313
	2	0.0515	0.1488	0.2376	0.2605	0.2936	0.3115	0.2965	0.2731	0.2587	0.2090	0.1569	0.1094
	3	0.0054	0.0331	0.0839	0.1042	0.1468	0.2076	0.2541	0.2731	0.2786	0.2787	0.2568	0.2188
	4	0.0004	0.0046	0.0185	0.0260	0.0459	0.0865	0.1361	0.1707	0.1875	0.2322	0.2627	0.2734
	5		0.0004	0.0026	0.0042	0.0092	0.0231	0.0467	0.0683	0.0808	0.1239	0.1719	0.2188
	6			0.0002	0.0004	0.0011	0.0038	0.0100	0.0171	0.0217	0.0413	0.0703	0.1094
	7					0.0001	0.0004	0.0012	0.0024	0.0033	0.0079	0.0164	0.0313
	8							0.0001	0.0002	0.0002	0.0007	0.0017	0.0039
9	0	0.6302	0.3874	0.2316	0.1938	0.1342	0.0751	0.0404	0.0260	0.0207	0.0101	0.0046	0.0020
	1	0.2985	0.3874	0.3679	0.3489	0.3020	0.2253	0.1556	0.1171	0.1004	0.0605	0.0339	0.0176
	2	0.0629	0.1722	0.2597	0.2791	0.3020	0.3003	0.2668	0.2341	0.2162	0.1612	0.1110	0.0703
	3	0.0077	0.0446	0.1069	0.1302	0.1762	0.2336	0.2668	0.2731	0.2716	0.2508	0.2119	0.1641
	4	0.0006	0.0074	0.0283	0.0391	0.0661	0.1168	0.1715	0.2048	0.2194	0.2508	0.2600	0.2461
	5		0.0008	0.0050	0.0078	0.0165	0.0389	0.0735	0.1024	0.1181	0.1672	0.2128	0.2461
	6		0.0001	0.0006	0.0010	0.0028	0.0087	0.0210	0.0341	0.0424	0.0743	0.1160	0.1641
	7			0.0001	0.0003	0.0012	0.0039	0.0073	0.0098	0.0212	0.0407	0.0703
	8					0.0001	0.0004	0.0009	0.0013	0.0035	0.0083	0.0176
	9							0.00001	0.0001	0.0003	0.0008	0.0020
10	0	0.5987	0.3487	0.1969	0.1615	0.1074	0.0563	0.0282	0.0173	0.0135	0.0060	0.0025	0.0010
	1	0.3151	0.3874	0.3474	0.3230	0.2684	0.1877	0.1211	0.0867	0.0725	0.0403	0.0207	0.0098
	2	0.0746	0.1937	0.2759	0.2907	0.3020	0.2816	0.2335	0.1951	0.1757	0.1209	0.0763	0.0439
	3	0.0105	0.0574	0.1298	0.1550	0.2013	0.2503	0.2668	0.2601	0.2522	0.2150	0.1665	0.1172
	4	0.0010	0.0112	0.0401	0.0543	0.0881	0.1460	0.2001	0.2276	0.2377	0.2508	0.2384	0.2051
	5	0.0001	0.0015	0.0085	0.0130	0.0264	0.0584	0.1029	0.1366	0.1536	0.2007	0.2340	0.2461
	6		0.0001	0.0012	0.0022	0.0055	0.0162	0.0368	0.0569	0.0689	0.1115	0.1596	0.2051
	7			0.0001	0.0002	0.0008	0.0031	0.0090	0.0163	0.0212	0.0425	0.0746	0.1172
	8					0.0001	0.0004	0.0014	0.0030	0.0043	0.0106	0.0229	0.0439
	9							0.0001	0.0003	0.0005	0.0016	0.0042	0.0098
	10		(all other entries < 0.0001)								0.0001	0.0003	0.0010

 ISBN: 9780170446938

Poisson Distribution

Each entry gives the probability that a Poisson random variable X, with the parameter λ, has the value x.

$$P(X = x) = \frac{\lambda^x e^{-\lambda}}{x!}$$
$$\mu = \lambda, \qquad \sigma = \sqrt{\lambda}$$

x \ λ	0.1	0.2	0.3	0.4	0.5	0.6	0.7	0.8	0.9	1.0
0	0.9048	0.8187	0.7408	0.6703	0.6065	0.5488	0.4966	0.4493	0.4066	0.3679
1	0.0905	0.1637	0.2222	0.2681	0.3033	0.3293	0.3476	0.3595	0.3659	0.3679
2	0.0045	0.0164	0.0333	0.0536	0.0758	0.0988	0.1217	0.1438	0.1647	0.1839
3	0.0002	0.0011	0.0033	0.0072	0.0126	0.0198	0.0284	0.0383	0.0494	0.0613
4		0.0001	0.0003	0.0007	0.0016	0.0030	0.0050	0.0077	0.0111	0.0153
5				0.0001	0.0002	0.0004	0.0007	0.0012	0.0020	0.0031
6							0.0001	0.0002	0.0003	0.0005
7										0.0001

x \ λ	1.1	1.2	1.3	1.4	1.5	1.6	1.7	1.8	1.9	2.0
0	0.3329	0.3012	0.2725	0.2466	0.2231	0.2019	0.1827	0.1653	0.1496	0.1353
1	0.3662	0.3614	0.3543	0.3452	0.3347	0.3230	0.3106	0.2975	0.2842	0.2707
2	0.2014	0.2169	0.2303	0.2417	0.2510	0.2584	0.2640	0.2678	0.2700	0.2707
3	0.0738	0.0867	0.0998	0.1128	0.1255	0.1378	0.1496	0.1607	0.1710	0.1804
4	0.0203	0.0260	0.0324	0.0395	0.0471	0.0551	0.0636	0.0723	0.0812	0.0902
5	0.0045	0.0062	0.0084	0.0111	0.0141	0.0176	0.0216	0.0260	0.0309	0.0361
6	0.0008	0.0012	0.0018	0.0026	0.0035	0.0047	0.0061	0.0078	0.0098	0.0120
7	0.0001	0.0002	0.0003	0.0005	0.0008	0.0011	0.0015	0.0020	0.0027	0.0034
8			0.0001	0.0001	0.0001	0.0002	0.0003	0.0005	0.0006	0.0009
9							0.0001	0.0001	0.0001	0.0002

x \ λ	2.2	2.4	2.6	2.8	3.0	3.2	3.4	3.6	3.8	4.0
0	0.1108	0.0907	0.0743	0.0608	0.0498	0.0408	0.0334	0.0273	0.0224	0.0183
1	0.2438	0.2177	0.1931	0.1703	0.1494	0.1304	0.1135	0.0984	0.0850	0.0733
2	0.2681	0.2613	0.2510	0.2384	0.2240	0.2087	0.1929	0.1771	0.1615	0.1465
3	0.1966	0.2090	0.2176	0.2225	0.2240	0.2226	0.2186	0.2125	0.2046	0.1954
4	0.1082	0.1254	0.1414	0.1557	0.1680	0.1781	0.1858	0.1912	0.1944	0.1954
5	0.0476	0.0602	0.0735	0.0872	0.1008	0.1140	0.1264	0.1377	0.1477	0.1563
6	0.0174	0.0241	0.0319	0.0407	0.0504	0.0608	0.0716	0.0826	0.0936	0.1042
7	0.0055	0.0083	0.0118	0.0163	0.0216	0.0278	0.0348	0.0425	0.0508	0.0595
8	0.0015	0.0025	0.0038	0.0057	0.0081	0.0111	0.0148	0.0191	0.0241	0.0298
9	0.0004	0.0007	0.0011	0.0018	0.0027	0.0040	0.0056	0.0076	0.0102	0.0132
10	0.0001	0.0002	0.0003	0.0005	0.0008	0.0013	0.0019	0.0028	0.0039	0.0053
11			0.0001	0.0001	0.0002	0.0004	0.0006	0.0009	0.0013	0.0019
12					0.0001	0.0001	0.0002	0.0003	0.0004	0.0006
13								0.0001	0.0001	0.0002
14										0.0001

x \ λ	4.2	4.4	4.6	4.8	5.0	5.2	5.4	5.6	5.8	6.0
0	0.0150	0.0123	0.0101	0.0082	0.0067	0.0055	0.0045	0.0037	0.0030	0.0025
1	0.0630	0.0540	0.0462	0.0395	0.0337	0.0287	0.0244	0.0207	0.0176	0.0149
2	0.1323	0.1188	0.1063	0.0948	0.0842	0.0746	0.0659	0.0580	0.0509	0.0446
3	0.1852	0.1743	0.1631	0.1517	0.1404	0.1293	0.1185	0.1082	0.0985	0.0892
4	0.1944	0.1917	0.1875	0.1820	0.1755	0.1681	0.1600	0.1515	0.1428	0.1339
5	0.1633	0.1687	0.1725	0.1747	0.1755	0.1748	0.1728	0.1697	0.1656	0.1606
6	0.1143	0.1237	0.1323	0.1398	0.1462	0.1515	0.1555	0.1584	0.1601	0.1606
7	0.0686	0.0778	0.0869	0.0959	0.1044	0.1125	0.1200	0.1267	0.1326	0.1377
8	0.0360	0.0428	0.0500	0.0575	0.0653	0.0731	0.0810	0.0887	0.0962	0.1033
9	0.0168	0.0209	0.0255	0.0307	0.0363	0.0423	0.0486	0.0552	0.0620	0.0688
10	0.0071	0.0092	0.0118	0.0147	0.0181	0.0220	0.0262	0.0309	0.0359	0.0413
11	0.0027	0.0037	0.0049	0.0064	0.0082	0.0104	0.0129	0.0157	0.0190	0.0225
12	0.0009	0.0013	0.0019	0.0026	0.0034	0.0045	0.0058	0.0073	0.0092	0.0113
13	0.0003	0.0005	0.0007	0.0009	0.0013	0.0018	0.0024	00032	0.0041	0.0052
14	0.0001	0.0001	0.0002	0.0003	0.0005	0.0007	0.0009	0.0013	0.0017	0.0022
15		0.0001	0.0001	0.0001	0.0002	0.0002	0.0003	0.0005	0.0007	0.0009
16						0.0001	0.0001	0.0002	0.0002	0.0003
17		(all other entries < 0.0001)						0.0001	0.0001	0.0001

ISBN: 9780170389389

Standard Normal Distribution

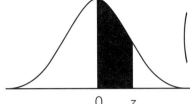

$$\left(Z = \frac{x - \mu}{\sigma} \right)$$

Each entry gives the probability that the standardised normal random variable Z lies between 0 and z.

Differences

z	0	1	2	3	4	5	6	7	8	9	1	2	3	4	5	6	7	8	9
0.0	.0000	.0040	.0080	.0120	.0160	.0199	.0239	.0279	.0319	.0359	4	8	12	16	20	24	28	32	36
0.1	.0398	.0438	.0478	.0517	.0557	.0596	.0636	.0675	.0714	.0754	4	8	12	16	20	24	28	32	36
0.2	.0793	.0832	.0871	.0910	.0948	.0987	.1026	.1064	.1103	.1141	4	8	12	15	19	22	27	31	35
0.3	.1179	.1217	.1255	.1293	.1331	.1368	.1406	.1443	.1480	.1517	4	8	11	15	19	22	26	30	34
0.4	.1554	.1591	.1628	.1664	.1700	.1736	.1772	.1808	.1844	.1879	4	7	11	14	18	22	25	29	32
0.5	.1915	.1950	.1985	.2019	.2054	.2088	.2123	.2157	.2190	.2224	3	7	10	14	17	21	24	27	31
0.6	.2258	.2291	.2324	.2357	.2389	.2422	.2454	.2486	.2518	.2549	3	6	10	13	16	19	23	26	29
0.7	.2580	.2612	.2642	.2673	.2704	.2734	.2764	.2794	.2823	.2852	3	6	9	12	15	18	21	24	27
0.8	.2881	.2910	.2939	.2967	.2996	.3023	.3051	.3078	.3106	.3133	3	6	8	11	14	17	19	22	25
0.9	.3159	.3186	.3212	.3238	.3264	.3289	.3315	.3340	.3365	.3389	3	5	8	10	13	15	18	20	23
1.0	.3413	.3438	.3461	.3485	.3508	.3531	.3554	.3577	.3599	.3621	2	5	7	9	12	14	16	18	21
1.1	.3643	.3665	.3686	.3708	.3729	.3749	.3770	.3790	.3810	.3830	2	4	6	8	10	12	14	16	19
1.2	.3849	.3869	.3888	.3907	.3925	.3944	.3962	.3980	.3997	.4015	2	4	5	7	9	11	13	15	16
1.3	.4032	.4049	.4066	.4082	.4099	.4115	.4131	.4147	.4162	.4177	2	3	5	6	8	10	11	13	14
1.4	.4192	.4207	.4222	.4236	.4251	.4265	.4279	.4292	.4306	.4319	1	3	4	6	7	8	10	11	13
1.5	.4332	.4345	.4357	.4370	.4382	.4394	.4406	.4418	.4429	.4441	1	2	4	5	6	7	8	10	11
1.6	.4452	.4463	.4474	.4484	.4495	.4505	.4515	.4525	.4535	.4545	1	2	3	4	5	6	7	8	9
1.7	.4554	.4564	.4573	.4582	.4591	.4599	.4608	.4616	.4625	.4633	1	2	3	3	4	5	6	7	8
1.8	.4641	.4649	.4656	.4664	.4671	.4678	.4686	.4693	.4699	.4706	1	1	2	3	4	4	5	6	6
1.9	.4713	.4719	.4726	.4732	.4738	.4744	.4750	.4756	.4761	.4767	1	1	2	2	3	4	4	5	5
2.0	.4772	.4778	.4783	.4788	.4793	.4798	.4803	.4808	.4812	.4817	0	1	1	2	2	3	3	4	4
2.1	.4821	.4826	.4830	.4834	.4838	.4842	.4846	.4850	.4854	.4857	0	1	1	2	2	2	3	3	4
2.2	.4861	.4864	.4868	.4871	.4875	.4878	.4881	.4884	.4887	.4890	0	1	1	1	2	2	2	3	3
2.3	.4893	.4896	.4898	.4901	.4904	.4906	.4909	.4911	.4913	.4916	0	0	1	1	1	2	2	2	2
2.4	.4918	.4920	.4922	.4925	.4927	.4929	.4931	.4932	.4934	.4936	0	0	1	1	1	1	1	2	2
2.5	.4938	.4940	.4941	.4943	.4945	.4946	.4948	.4949	.4951	.4952	0	0	0	1	1	1	1	1	1
2.6	.4953	.4955	.4956	.4957	.4959	.4960	.4961	.4962	.4963	.4964	0	0	0	0	1	1	1	1	1
2.7	.4965	.4966	.4967	.4968	.4969	.4970	.4971	.4972	.4973	.4974	0	0	0	0	0	1	1	1	1
2.8	.4974	.4975	.4976	.4977	.4977	.4978	.4979	.4979	.4980	.4981	0	0	0	0	0	0	0	0	1
2.9	.4981	.4982	.4982	.4983	.4984	.4984	.4985	.4985	.4986	.4986	0	0	0	0	0	0	0	0	1
3.0	.4987	.4987	.4987	.4988	.4988	.4989	.4989	.4989	.4990	.4990	0	0	0	0	0	0	0	0	0
3.1	.4990	.4991	.4991	.4991	.4992	.4992	.4992	.4992	.4993	.4993	0	0	0	0	0	0	0	0	0
3.2	.4993	.4993	.4994	.4994	.4994	.4994	.4994	.4995	.4995	.4995	0	0	0	0	0	0	0	0	0
3.3	.4995	.4995	.4995	.4996	.4996	.4996	.4996	.4996	.4996	.4997	0	0	0	0	0	0	0	0	0
3.4	.4997	.4997	.4997	.4997	.4997	.4997	.4997	.4997	.4998	.4998	0	0	0	0	0	0	0	0	0
3.5	.4998	.4998	.4998	.4998	.4998	.4998	.4998	.4998	.4998	.4998	0	0	0	0	0	0	0	0	0
3.6	.4998	.4998	.4999	.4999	.4999	.4999	.4999	.4999	.4999	.4999	0	0	0	0	0	0	0	0	0
3.7	.4999	.4999	.4999	.4999	.4999	.4999	.4999	.4999	.4999	.4999	0	0	0	0	0	0	0	0	0
3.8	.4999	.4999	.4999	.4999	.4999	.4999	.4999	.5000	.5000	.5000	0	0	0	0	0	0	0	0	0
3.9	.5000	.5000	.5000	.5000	.5000	.5000	.5000	.5000	.5000	.5000	0	0	0	0	0	0	0	0	0

 ISBN: 9780170446938

Instructions

1. Pull this page out of your book and cut along the lines.

2. For each distribution find three boxes with bolded teal script.

3. Find one box showing a possible graph for each distribution.

4. Match each of the problem boxes (A to J) with the correct distribution.

5. Select an appropriate model for each problem, list the parameters and then use your model to solve the problem.

Poisson

Discrete distribution

Parameter: λ

Rare events

$P(X = x)$ vs x (0 to 7)

A At Paradise High School, the mean number of false fire alarms each term is 1.2. What is the probability that there are no false fire alarms in a term?

H The school chess team loses, on average, 3.1 games each term. Calculate the probability that they will lose no fewer than five games in a term.

Binomial

Discrete distribution

Parameters: n and π

Fixed number of trials, each with two outcomes

$P(X = x)$ vs x (0 to 10)

D The boys' netball team has a probability of 0.55 of winning a game. If they play 12 games, calculate the probability that they win more than eight.

G A striker in the football team scores a goal, on average, in 18% of attempts. If he makes seven attempts during a game, calculate the probability that he is successful in at least three.

Normal	Rectangular	Triangular
Continuous distribution	Continuous distribution	Continuous distribution
Parameters: μ and σ	Parameters: a and b	Parameters: a, b and c
Symmetrical bell-shaped curve	Has maximum and minimum but no mode	Has maximum and minimum and mode

F The mean distance between the pupils of some students' eyes of is found to be 53 mm with a standard deviation of 3 mm. Calculate the probability that the distance between the eyes of a student is more than 55 mm.

I Buses arrive at a bus stop at six-minute intervals. Assuming that people arrive at the stop at random intervals, calculate the probability that a person will have to wait for longer than four and a half minutes.

B The time taken to complete a crossword ranged from 2 minutes to 20 minutes, with the most common time being 15 minutes. Calculate the probability that a person took more than 10 minutes to complete the puzzle.

C A flock of lambs weigh, on average, 17.4 kg, with a standard deviation of 0.6 kg. What proportion of the flock weighs between 17 and 18 kg?

J A farmer's sheep have escaped, so he knows there is a hole in his fence. It is 270 m long, and he can see that there is no hole in the first 20 m. Calculate the probability that the hole is within the first 100 m.

E Students at Paradise High School live within 12 km of the school. The most common distance from the school is 2 km. Calculate the probability that a student lives between one and two kilometres from the school.